Bridge Structural Health Monitoring

Advances in predictive maintenance and smart technology enable structural health monitoring (SHM) of critical infrastructure like bridges. Online SHM promises to optimize maintenance, extend lifespan, and ensure safety and reliability. This practitioner reference and textbook focuses on the use of SHM in a truss bridge management system, comparing calculated displacements, strains, and accelerations with calculated results. It emphasizes the use of SHM data for predictive asset management. It outlines the instrumentation techniques and field loading testing for static and dynamic evaluation, pre-processing of raw data, theoretical and experimental modal analysis, indicators sensitive to structural damage, and the validation of numerical models. It introduces a decision support system for structure maintenance management, conducting a capacity assessment of structural members using a detailed steel truss railway bridge case study.

This book is an essential introductory book for undergraduate structural engineering students while detailed enough to introduce SHM to practicing engineers.

Shervan Ataei serves as a faculty member at the School of Railway Engineering, Iran University of Science and Technology since 2006. He teaches courses in bridge design, maintenance of railway structures, and structural health monitoring of railway infrastructure. Dr. Ataei earned his PhD in structural engineering from Tarbiat Modares University.

Amin Miri is a structural engineer specialising in railway infrastructure and structural health monitoring. He has completed a PhD in structural engineering at Queensland University of Technology, receiving a Higher Achiever Award for his outstanding research publications. Dr. Miri is a registered Professional Engineer and a member of Engineers Australia.

Bridge Structural Health Monitoring

The Truss Bridge Management System

Shervan Ataei and Amin Miri

CRC Press
Taylor & Francis Group
Boca Raton London New York

CRC Press is an imprint of the
Taylor & Francis Group, an **informa** business

Designed cover image: Shervan Ataei and Amin Miri

First edition published 2026
by CRC Press
4 Park Square, Milton Park, Abingdon, Oxon, OX14 4RN

and by CRC Press
2385 NW Executive Center Drive, Suite 320, Boca Raton FL 33431

CRC Press is an imprint of Informa UK Limited

British Library Cataloguing-in-Publication Data
A catalogue record for this book is available from the British Library

ISBN: 9781032789378 (hbk)
ISBN: 9781032799452 (pbk)
ISBN: 9781003494652 (ebk)

DOI: 10.1201/9781003494652

Typeset in Times
by Newgen Publishing UK

Dedication

Gratitude to God, who creates, guides, ordains death, and grants life.
Dedicated to the steadfast pillars of my life, my beloved parents.
Dedicated to my companion in life, my esteemed and learned spouse.
Dedicated to the divine trusts, my cherished children.
Dedicated also to Fereydoun, Ahmad, Amirhossein, Abdolhamid, and Mehdi, who fulfilled their knowledge and their covenant with God.
Dedicated to Saeed, whose support was instrumental in the creation of the knowledge embodied in this work.
With thanks to Amin, whose collaboration greatly aided the writing of this book.

Contents

Preface

Structural design is often regarded as the pinnacle of an engineering curriculum, where the essence of all coursework converges. To account for uncertainties during construction and operation, conventional design relies on conservative safety factors. However, advances in structural monitoring now allow for measurement-based design—before, during, and after construction—enabling the use of reduced safety factors without compromising reliability and leading to more optimized structures. Thus, structural monitoring has the potential to transform design practice.

Field measurements also provide an invaluable opportunity to compare structural response with theoretical predictions taught in core engineering curriculum such as statics, structural analysis, strength of materials, matrix methods, structural dynamics, and finite element analysis. This integration allows a deeper, experiential understanding of engineering principles.

Maintenance and repair represent some of the most critical and costly stages in a structure's lifecycle. Current maintenance strategies are often reactive—repairs are made after damage occurs—or based on fixed schedules. To reduce lifecycle costs and transition toward condition-based or predictive maintenance, advanced field measurements of structural response are essential. In this sense, structural monitoring can revolutionise maintenance practices.

The dual imperative of ensuring safety and efficiency, on the one hand, and advancing technologies, on the other, has placed structural health monitoring among the most important research areas worldwide. Bridges, as vital components of road and rail infrastructure, are at the forefront of this field.

This book introduces the fundamentals of structural health monitoring through a self-study, project-based approach. Core concepts are presented and implemented in the context of benchmark field experiments, equipping readers with the knowledge and skills required to monitor structures in general, and bridges in particular. A broad, step-by-step perspective guides the reader toward practical bridge-monitoring projects. The book's educational goal is to present practical methods for measurement and data analysis. For more advanced techniques, readers are directed to specialized references.

The book is organized into seven chapters:

- Chapter 1 presents a six-stage maintenance model based on ISO 55000 and clarifies the role of structural monitoring in advancing toward condition-based maintenance.
- Chapter 2 demonstrates visual inspection of a truss bridge as well as instrumented inspections of road and railway bridges.
- Chapter 3 introduces advanced monitoring tools—strain gauges, displacement sensors, and accelerometers—commonly used to capture global structural response of structures.

- Chapter 4 discusses load rating with and without field load testing, illustrated by static and dynamic load tests on a truss bridge. The chapter also covers signal acquisition and essential preprocessing steps such as baseline correction, filtering, and resampling using MATLAB functions.
- Chapter 5 presents modal analysis, both theoretical (for multi-degree-of-freedom systems) and experimental (using free-vibration data), applied to the truss bridge case study.
- Chapter 6 introduces well-known damage-sensitive indicators, highlighting the importance of eliminating environmental and operational effects so that variations in the indicators reflect only structural damage. Applications include damage detection and service-life estimation.
- Chapter 7 discusses service-life planning as a key application of structural monitoring, along with methods for assessing reserve capacity and actual loading demands, illustrated with bridge and railway case studies.

The book is suitable for undergraduate and graduate courses in structural monitoring within civil engineering courses.

The authors gratefully acknowledge the contributions of students at the Railway Engineering Department, Iran University of Science and Technology, over the past decade. As with all human endeavours, this book is not free from error, and the authors welcome feedback and suggestions at **ataei@iust.ac.ir** for future editions.

A companion website offering accessible resources has been provided: https://www.routledge.com/9781032789378

Autumn 2025

About the Authors

Shervan Ataei was born in Tehran. He received his Bachelor's degree in Civil Engineering in 1998 from the Faculty of Engineering at the University of Tehran. In 2000, he collaborated with his supervisor, Dr. Marefat, on the load testing of masonry arch bridges at kilometres 23 and 24 of the old Tehran–Qom railway line. That same year, he earned his Master's degree in Structural Engineering from the University of Tehran.

In 2001, he worked with Dr. Marefat on the dynamic load testing of the Tele Zang railway bridge. In 2002, he participated in the field load testing of the Neka Bridge, and in 2004, in the load testing of the Ghal'e Morghi Bridge and tracks 4 and 5 of Bahram Station in collaboration with Dr. Mohammadzadeh. His doctoral dissertation, completed in 2006 under the supervision of Dr. Aghakochak, was titled "Vibration-Based Condition Assessment of Railway Bridges Using Measured Responses," earning him a PhD in Structural Engineering from Tarbiat Modares University.

Since 2006, he has served as a faculty member at the School of Railway Engineering, Iran University of Science and Technology, where he teaches courses in bridge design, maintenance of railway structures, and structural health monitoring of railway infrastructure. Through industrial projects, he has conducted field load testing on ten masonry arch bridges and two steel railway bridges. In addition, he has carried out long-term monitoring of one masonry bridge, two steel bridges, one prestressed concrete bridge, and a railway turnout under service loads. He has also developed a dynamic weighing system for axle loads at a railway station.

Amin Miri is a structural engineer specialising in railway infrastructure and structural health monitoring. He holds a Bachelor's degree in Track and Infrastructure Engineering from the Railway Engineering Department at Iran University of Science and Technology, where he graduated as the top student of his cohort. He later completed a PhD in Structural Engineering at the Queensland University of Technology, receiving a Higher Achiever Award for his outstanding research publications.

Over the course of nearly a decade, Amin has worked extensively in the railway sector, including roles with Iranian Railways, Queensland Rail, and Rio Tinto, where he is currently employed as a Track Infrastructure Engineer. His professional and research efforts have been deeply focused on the performance and monitoring of bridges under modern railway demands. He has been directly involved in the structural health monitoring of multiple bridges, both in Iran and Australia, gaining practical expertise in assessing and maintaining critical transport assets.

Amin's research has led to more than 25 publications in peer-reviewed journals, particularly in the fields of structural health monitoring and railway engineering. His early work at Iranian Railways centred on the challenging task of evaluating the response of masonry arch bridges under increasing loads, in the absence of established standards. This project contributed to the development of new approaches for assessing such structures and ultimately inspired the creation of this book, which aims to consolidate knowledge and provide practical guidance for engineers and researchers working in this area.

Amin is a registered professional engineer and a member of Engineers Australia. He has also contributed to academic development through student supervision during his doctoral research. His combined academic achievements, professional experience, and practical insights form the foundation of his continued contributions to advancing knowledge in railway engineering and structural health monitoring.

1 Condition-Based Maintenance

1.1 INTRODUCTION

Maintenance plays a crucial role in ensuring the longevity and functionality of assets across various sectors. According to the definition provided by EN13306, maintenance is a combination of managerial and technical activities performed throughout the life cycle of an asset to maintain or restore it to a condition where it can perform its expected functions [1]. Assets can be broadly categorized into human, financial, informational, intangible, and physical assets. This book primarily focuses on the condition-based maintenance (CBM) of physical assets within transportation infrastructure, which often represents the backbone of economic growth and societal mobility.

Physical assets, particularly in transportation infrastructure, represent significant investments. These assets, including bridges, tunnels, railway tracks, and highways, form the core of a nation's connectivity and commerce. Given their substantial value, even a modest reduction in maintenance expenses can lead to substantial savings and improved asset performance. As an illustrative example, the road transport infrastructure in Iran was valued at approximately 3,000 trillion Iranian rials in 2021 (equivalent to around three quadrillion IRR). Maintaining this infrastructure requires an annual budget of about 5% of this value (roughly 150 trillion IRR) for routine maintenance and repairs. This sizeable investment highlights the importance of employing effective maintenance strategies.

In recent years, the adoption of CBM has gained momentum as a more efficient and cost-effective approach. Unlike traditional preventive maintenance, which is performed based on predetermined schedules, CBM involves continuous monitoring of asset conditions, allowing maintenance to be conducted only when necessary. This approach not only minimizes downtime but also extends the life of assets, reducing overall lifecycle costs. A study conducted by Garg and Deshmukh [2] indicated that proactive CBM could extend the life of assets by up to 25%, thereby substantially lowering the likelihood of unexpected failures and catastrophic events.

Integrating CBM within transportation infrastructure requires a systematic approach that combines engineering methodologies with economic principles. This

DOI: 10.1201/9781003494652-1

integration leads to more informed decision-making and enables the efficient allocation of resources. A critical guideline in this context is the ISO 55000 standard [3], which provides a comprehensive framework for asset management and helps organizations optimize the value derived from their assets.

1.2 ISO 55000 STANDARD

The ISO 55000 [3] standard serves as an international benchmark for asset management, outlining the fundamental requirements for establishing, implementing, maintaining, and improving an effective asset management system. It emphasizes leadership, planning, support, operations, performance evaluation, and continuous improvement. The core objective of ISO 55000 is to ensure that asset management activities are aligned with organizational objectives, resulting in reduced failures, enhanced safety, improved reliability, increased availability, minimized downtime, and lower maintenance costs.

Implementing ISO 55000 fosters a structured approach to asset maintenance management, enabling organizations to develop clear strategies for improving performance, managing risks, and optimizing lifecycle costs. This systematic approach ensures that assets are maintained to meet the strategic objectives of the organization, contributing to their long-term sustainability and effectiveness [3].

A notable example of ISO 55000 implementation can be seen with the Sydney Harbour Bridge in Australia [4]. Constructed in the 1930s, this iconic steel arch bridge represents a critical transportation asset that required effective maintenance management to ensure its long-term structural integrity while minimizing disruptions to daily traffic. In 2016, Transport for New South Wales adopted the ISO 55000 standard to enhance the bridge's maintenance strategy [4, 5].

The implementation involved developing a comprehensive asset management plan that defined inspection intervals, maintenance strategies, and risk management procedures. Condition monitoring techniques, such as ultrasonic testing and visual inspections, were used to detect early signs of fatigue and corrosion [5]. Additionally, a digital asset management system was introduced, enabling real-time data collection and analysis, which facilitated informed decision-making regarding maintenance schedules [6].

As a result, maintenance costs for the Sydney Harbour Bridge were reduced by approximately 20% over a five-year period, while the asset's availability increased due to the scheduling of maintenance activities during off-peak hours [7]. This successful application of ISO 55000 highlights how proactive and systematic asset management can lead to substantial cost savings and improved asset performance in transportation infrastructure.

Numerous researchers have developed models for implementing ISO 55000 in different industries. These models provide a roadmap for organizations seeking to adopt best practices in asset management. A simplified version of one such model is discussed below, illustrating how the principles of ISO 55000 can be practically applied within transportation infrastructure.

1.3 THE MAINTENANCE MANAGEMENT EXCELLENCE MODEL

Improper maintenance is often the primary cause of asset failure, resulting in costly repairs, downtime, and even catastrophic events. Common causes include design flaws, inappropriate operation, insufficient training, poor-quality materials, inadequate supervision, non-compliance with repair standards, and failure to adhere to operational guidelines [8]. To achieve excellence in maintenance management for infrastructure assets, six essential steps must be undertaken and repeated in a cyclical manner, with continuous refinement and enhancement to improve the model's output (Figure 1.1). This process aligns with the Plan–Do–Check–Act (PDCA) cycle, also known as the Deming cycle [9].

The six stages of this model involve:

1. Cultural transformation for asset maintenance
2. Asset information management
3. Preventing premature deterioration of assets
4. Optimizing maintenance programmes
5. Lifecycle management activities
6. Performance management and continuous improvement

The following sections will delve into each of these stages, demonstrating their significance in achieving maintenance excellence.

1.3.1 Cultural Transformation for Asset Maintenance

The first stage of the maintenance excellence model is fostering a cultural transformation within the organization regarding asset maintenance. Achieving maintenance excellence is not solely dependent on technical measures but also on the mindset and behaviour of the people involved. A shift in attitudes among managers, employees, and asset operators is crucial to promote a proactive approach toward maintenance activities rather than a reactive one.

Organizational leadership plays a pivotal role in driving this cultural shift. The commitment of top management signals the importance of maintenance excellence to the entire workforce. Research has shown that leadership commitment is one of the strongest predictors of the successful implementation of maintenance strategies. This commitment should manifest in terms of resource allocation, priority setting, and the creation of an inclusive environment where all employees feel responsible for asset health.

To ensure effective coordination and accountability, an individual responsible for implementing maintenance excellence should be appointed by the organization's management. This individual, often referred to as a "Maintenance Excellence Champion" or "Reliability Leader," is tasked with overseeing maintenance initiatives, tracking performance metrics, and communicating the strategic importance of maintenance excellence across departments.

This decision should be communicated to all staff to ensure that everyone understands the focus on maintenance as a strategic priority. The designation of a leader emphasizes that maintenance is not just the responsibility of the technical team but a core organizational objective.

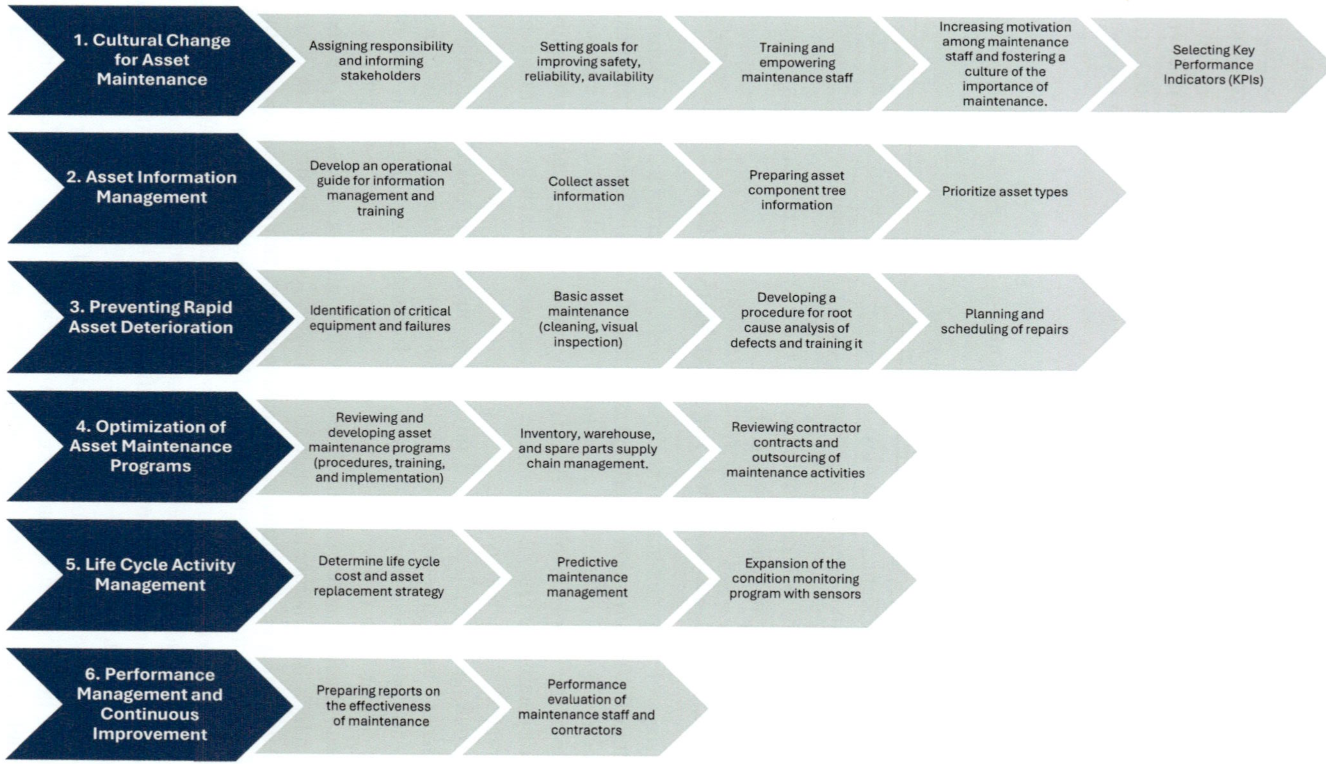

FIGURE 1.1 Simplified maintenance excellence model.

Regular meetings are essential to foster collaboration and provide a platform for cross-functional discussions. During these meetings, management should work with the maintenance team to define strategies that align with technical, organizational, and financial objectives. These meetings are also an opportunity to establish and review key performance indicators (KPIs), such as:

- Mean time between failures (MTBF): A key indicator of the reliability of equipment.
- Mean time to repair (MTTR): A measure of maintenance effectiveness and speed.
- Overall equipment effectiveness (OEE): A comprehensive measure that evaluates the performance of equipment based on availability, performance, and quality.

These KPIs must be directly tied to business goals such as minimizing downtime, reducing operational costs, and maximizing asset lifespan. Establishing KPIs should be complemented by regular reviews of the necessary actions to achieve these objectives, ensuring continuous alignment with broader business strategies.

Establishing a direct link between employee rewards and the outcomes of their maintenance efforts can serve as a strong motivator. Reward structures can be designed around the successful completion of maintenance tasks, the improvement of KPIs, or innovations in predictive maintenance approaches. This approach fosters a positive atmosphere of cooperation among managers, staff, and contractors, and encourages a shared responsibility for asset health.

Incentives should be tailored to recognize both individual and team contributions, as teamwork is often key in collaborative maintenance efforts. Research shows that organizations that have implemented reward systems linked to maintenance performance have reported improvements in staff engagement and overall operational efficiency [8].

To support the implementation of maintenance excellence, training programmes should be planned for both internal personnel and contractors. These programmes should focus not only on technical skills but also on enhancing understanding of maintenance best practices, reliability engineering principles, and the use of advanced maintenance technologies such as condition monitoring and predictive maintenance.

For example, vibration analysis and thermography are key components of predictive maintenance for rotating machinery. Training on how to interpret data from sensors and transform it into actionable insights is essential for effective decision-making. Additionally, guidelines should be developed for calculating evaluation indicators, analysing results, and preparing comprehensive maintenance reports. These guidelines will provide a structured approach to ensure consistency in evaluation and reporting.

Establishing a culture of continuous improvement is critical to the long-term success of maintenance excellence initiatives. Maintenance teams must not only focus on resolving current issues but also on identifying opportunities to prevent future problems. This can be achieved through the integration of lean maintenance principles, which focus on eliminating waste (e.g., downtime, excess inventory) and optimizing the use of resources.

Studies have shown that organizations with a strong maintenance culture can achieve up to a 30% increase in productivity [10]. Moreover, companies that foster a culture of continuous learning are more likely to innovate and adapt to new maintenance technologies such as the Industrial Internet of Things (IoT), which allows real-time monitoring of asset conditions and predictive analytics.

An example of successful cultural transformation can be observed in the case of the Tata Steel Plant in Jamshedpur, India. The company initiated a cultural change programme to improve maintenance practices across its facilities. By implementing regular training programmes, establishing a reward system linked to maintenance performance, and holding frequent maintenance strategy meetings, Tata Steel managed to reduce unplanned downtime by 25% within two years [11].

1.3.2 Asset Information Management

The second stage focuses on asset information management, which involves developing and disseminating asset management guidelines. This stage is crucial for ensuring that the assets are well-documented, their current conditions are understood, and their criticality within the broader infrastructure system is assessed. A sound asset information management system provides the foundation for effective maintenance, risk management, and performance optimization.

In large infrastructure networks, proper asset information management is vital for lifecycle management. This stage is not only about documenting assets but also about integrating various data sources to enable informed decision-making. Advanced technologies such as geographic information systems (GIS), building information modelling (BIM), and enterprise asset management (EAM) systems are frequently employed to support this process. These tools provide a comprehensive view of the asset portfolio, allowing stakeholders to visualize the spatial and operational relationships between assets.

A notable example of effective asset information management is the implementation of the asset registry at the London Underground. In 2016, Transport for London (TfL) invested in an asset information management system that centralized data on over 1,000 km of tracks, stations, and signalling systems [12]. This system leveraged data integration to enhance the management of one of the world's oldest and most complex underground rail networks. By implementing this system, TfL achieved a 15% reduction in maintenance costs due to improved asset tracking, scheduling, and condition monitoring.

The London Underground project highlights the importance of having a centralized repository of asset data. Previously, asset information was scattered across different departments, leading to inefficiencies in maintenance planning and execution. With the new system, real-time data on asset conditions, scheduled maintenance, and lifecycle performance became accessible across all levels of the organization. This case also underscores the value of predictive maintenance strategies enabled by such systems, where data analytics and condition monitoring techniques help predict when and where failures might occur, reducing unplanned downtime and emergency repairs [12].

In this phase, assets should be inventoried systematically, and their hierarchical structures and components should be established. This means creating a detailed catalogue of all infrastructure assets, from major components such as bridges, tunnels, and track segments, to smaller but equally critical parts like sensors, actuators, and control systems. Accurate and comprehensive data collection is key at this stage, as incomplete or inaccurate information can lead to poor decision-making down the line.

Assets should be grouped and classified based on functional, operational, and geographical criteria. A unique identification system is essential for tracking each asset throughout its lifecycle. For example, each piece of equipment should have an equipment-based code, while location-based codes should be used for where assets are installed. This enables seamless integration with maintenance schedules, operational activities, and condition monitoring programmes.

The ISO 14224 standard is highly relevant in this stage, offering guidelines for developing hierarchical structures and asset components [13]. This standard outlines the taxonomy and methodology for collecting and analysing data on equipment reliability and maintenance performance, thus improving the reliability of the asset management process.

The ISO 14224 standard [13] categorizes essential asset information into several components, each playing a critical role in the asset management process:

- *Basic information*: This includes the asset's identity, hierarchical structure, and technical specifications, which form the foundation of the asset registry.
- *Failure consequences*: This category deals with the potential consequences of asset failure, such as hidden, safety-related, environmental, performance, and infrastructure-related impacts. For example, the failure of a key bridge component could lead to significant safety risks, whereas the failure of a sensor might have only performance-related impacts.
- *Causes of failure*: Failure causes include design errors, construction mistakes, operational errors, and managerial faults. Understanding the root cause of failures is crucial for developing effective mitigation strategies. For example, operational errors could be mitigated by improving staff training, while design flaws might necessitate engineering reviews or redesigns.

Figure 1.2 illustrates how these failure causes manifest in real-world scenarios, showing a hierarchical breakdown of asset failures and their origins.

ISO 14224 also classifies failure mechanisms, including:

- *Mechanical failures*: These involve physical breakdowns such as leakage, vibration, or deformation. For example, in pipelines, excessive vibration can lead to fatigue and eventual failure at welded joints.
- *Material failures*: These include phenomena like corrosion, coating loss, cracks, fatigue, fire, and explosion. A notable example is the corrosion of steel

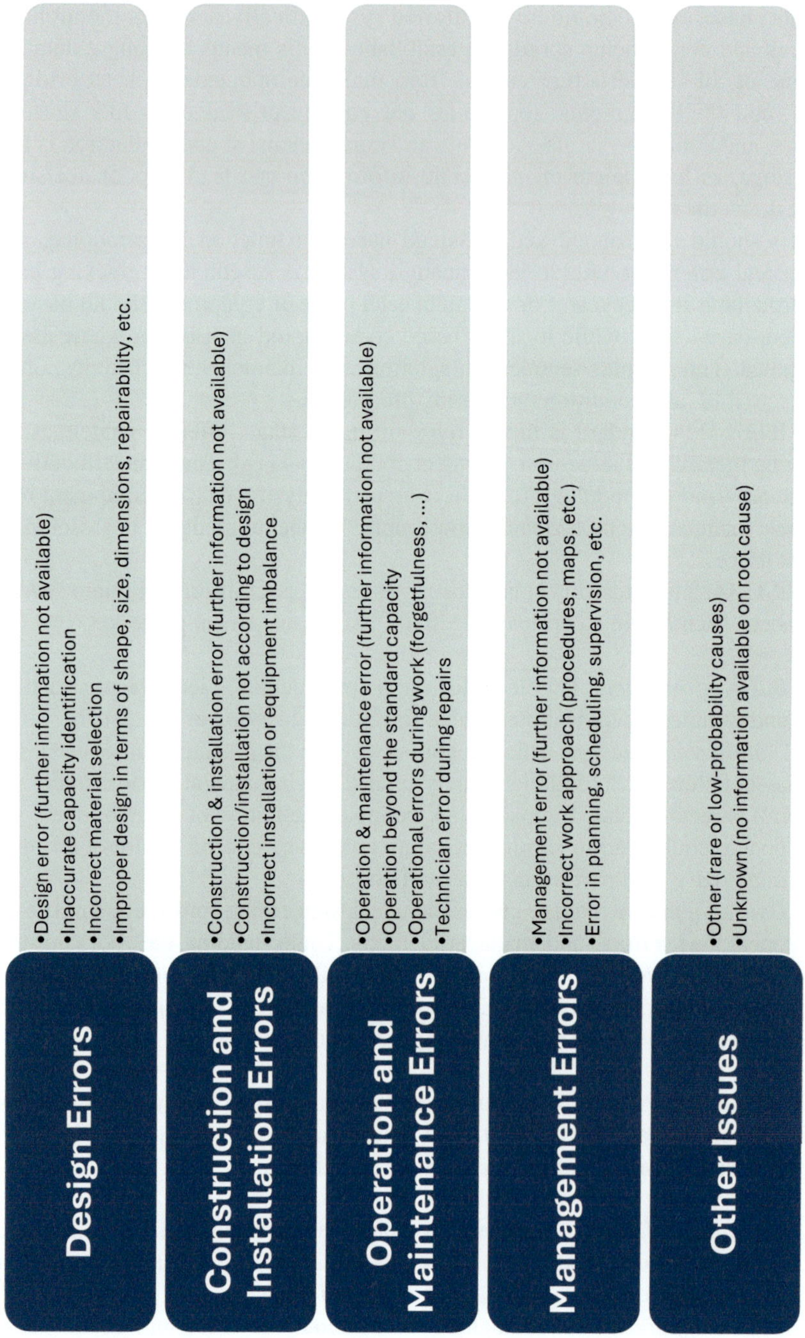

FIGURE 1.2 Causes of failure according to ISO 14224 [13].

bridges, where environmental exposure can lead to loss of structural integrity unless proper protective measures (such as coatings) are applied.

- *External failures*: These include issues such as sedimentation, freezing, or contamination, which often occur due to environmental factors. For instance, sedimentation in water pipes can reduce flow rates and cause pressure build-up, leading to pipe failure.

It also provides an overview of various detection methods that can be applied, such as visual inspections, ultrasonic testing, vibration analysis, and thermal imaging. In recent years, advanced technologies like drones and automated robots equipped with sensors have been used for remote condition assessments, improving safety and accuracy in detecting potential issues.

1.3.3 Preventing Premature Deterioration of Assets

In the third stage of the maintenance excellence model, efforts are made to prevent premature deterioration by addressing defects in critical assets. This stage focuses on ensuring that identified defects are rectified promptly, resulting in rapid improvements in asset performance. Critical assets are typically those that, if they fail, could lead to severe safety, environmental, or economic impacts. These assets are prioritized for intensive monitoring and maintenance efforts, following a structured approach such as a root cause analysis (RCA) programme [14].

RCA is essential in this stage, as it identifies underlying causes of recurrent issues rather than simply treating symptoms. By eliminating the root cause, the risk of future failures is minimized. RCA programmes often use tools like the "5 whys" or fishbone diagrams to systematically drill down into the problem and uncover its cause.

A vivid example of the importance of such maintenance approaches can be seen in the 2017 New York City Subway System crisis. Faced with frequent train delays and breakdowns, the subway system experienced disruptions that severely impacted its over 5 million daily riders. Upon investigation, it was discovered that the majority of these issues stemmed from deferred maintenance, ageing infrastructure, and recurring equipment failures.

In response, the Metropolitan Transportation Authority (MTA) adopted a comprehensive maintenance excellence strategy that prioritized preventive maintenance. Key elements of their programme included the deployment of rapid response teams, enhanced system-wide inspections, and regular performance evaluations. Through these efforts, the MTA achieved a 30% reduction in major incidents within just one year [15], demonstrating the transformative power of a well-implemented maintenance programme.

To successfully implement this stage of the programme, organizations must undertake the following actions:

- *Appoint a responsible individual*: Designating a dedicated individual or team to oversee the maintenance programme ensures accountability and effective

execution. This person is responsible for tracking progress, ensuring compliance with guidelines, and facilitating communication across departments.
- *Develop RCA guidelines*: Standardizing RCA procedures ensures consistency and thoroughness when investigating failures. These guidelines should include techniques for problem identification, data collection, and analysis to ensure all relevant factors are considered.
- *Train maintenance personnel, operators, and contractors*: Personnel must be equipped with the skills to identify potential defects and execute the necessary corrective actions. This requires regular training sessions focusing on the latest diagnostic technologies and preventive maintenance practices.
- *Create a database of failure causes*: This database serves as a reference for maintenance teams, helping them recognize patterns in equipment failures and implement more targeted maintenance strategies. Over time, it also contributes to the continuous improvement of maintenance practices by identifying recurring issues.
- *Evaluate the programme's effectiveness*: Regular assessments are crucial to ensure the maintenance programme is delivering expected results. KPIs, such as MTBF, MTTR, and OEE, should be monitored to gauge improvements.

During this stage, basic maintenance activities such as cleaning, tightening, lubrication, and visual inspections should be conducted systematically using checklists. These routine activities are vital to maintaining equipment in optimal condition and preventing minor defects from escalating into major failures. Additionally, *condition monitoring* plays a pivotal role in this phase, particularly for critical components.

Nondestructive testing (NDT) methods are employed to assess the health of assets without causing damage. Key NDT techniques include:

- *Vibration analysis*: Used to detect misalignments, unbalance, and bearing faults in rotating machinery.
- *Thermography*: A technique that utilizes infrared cameras to detect temperature variations, often indicative of electrical or mechanical problems such as overheating.
- *Ultrasonic testing*: High-frequency sound waves are used to detect internal flaws, such as cracks or cavity, in materials.
- *Eddy current testing*: This method is used to detect surface and near-surface defects in conductive materials.
- *Acoustic emission*: Sensors detect high-frequency waves emitted by the rapid release of energy from localized sources within a material, often indicating crack propagation or other structural defects.

These techniques allow for early detection of wear and damage, enabling timely interventions that can extend asset life and improve safety. Studies show that *condition-based maintenance* can reduce overall maintenance costs by 25% and minimize unplanned downtime by 45% [16].

During this phase, maintenance management processes should be optimized by:

- *Prioritizing repair requests*: High-priority issues, particularly those related to safety or critical assets, should be addressed promptly.
- *Planning and scheduling activities*: Effective planning ensures that resources (materials, labour, and time) are utilized efficiently. Scheduling preventive maintenance during planned downtimes or low production periods can help minimize disruptions.
- *Calculating maintenance efficiency indicators*: KPIs such as maintenance cost per unit output, maintenance backlog, and work order completion rate provide valuable insights into the efficiency of maintenance operations. Regular review of these metrics allows for continuous improvement in the maintenance process.

1.3.4 Optimizing Maintenance Programmes

Maintenance activities encompass a wide range of actions, such as replacement, repair, modification, adjustment, restoration, inspection, servicing, testing, comprehensive overhaul, or a combination of these activities. If these actions are performed after a failure has occurred, they fall under corrective maintenance (CM). This approach is reactive, where repairs or replacements are made only after failure has already disrupted the system. While CM can often be costly and lead to significant downtime, it is sometimes unavoidable, especially when equipment failure is unpredictable or difficult to monitor.

Conversely, if these actions are carried out pre-emptively, they are classified as preventive maintenance (PM). The goal of PM is to prevent failure before it occurs, thereby reducing unscheduled downtime and improving system reliability. Preventive maintenance activities are further categorized into two types: scheduled (time-based) and condition-based maintenance (CBM), as illustrated in Figure 1.3.

Scheduled preventive maintenance, also known as time-based maintenance, involves performing maintenance activities at predefined intervals, regardless of the equipment's condition. For example, routine servicing like cleaning, lubrication, and tightening, along with scheduled periodic replacements and performance testing, falls under this category. This type of maintenance is easy to plan and implement but may lead to unnecessary maintenance or premature component replacement, which can result in additional costs and resource use. An example of scheduled PM is the regular changing of oil in mechanical systems, even when the oil's condition may still be acceptable.

To complete, condition-based preventive maintenance (CBM) relies on the actual condition of the asset to decide when maintenance is required. By utilizing advanced monitoring tools such as vibration analysis, acoustic monitoring, thermal imaging, and oil analysis, CBM helps detect signs of deterioration and potential failures before they become critical. For instance, vibration analysis is widely used in rotating machinery, where abnormalities in vibration frequencies can indicate misalignment or bearing wear, signalling that maintenance is needed.

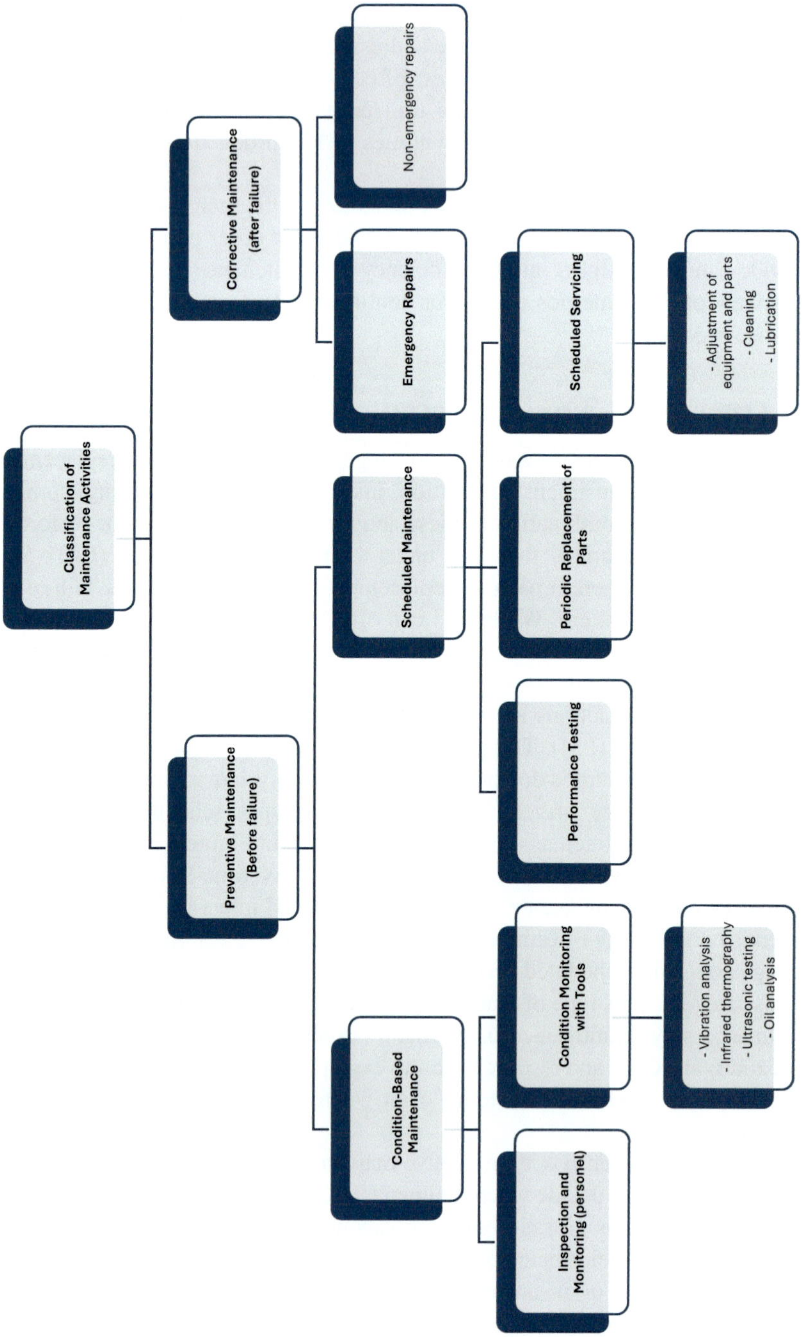

FIGURE 1.3 Classification of maintenance activities according to ISO14224 [13].

CBM is an essential element of SHM, particularly in fields such as civil infrastructure, aerospace, and energy sectors, where unexpected failures can have catastrophic consequences. For example, CBM in bridges involves using sensors to measure parameters like strain, displacement, and temperature, allowing engineers to detect structural weaknesses before they lead to failure.

In the fourth stage of the maintenance excellence programme, maintenance plans and resource allocation are optimized. The maintenance programmes for critical assets are reviewed using reliability-centred maintenance (RCM) methodologies. RCM is a systematic approach for determining the most cost-effective maintenance strategy, focusing on the functionality of critical assets and how they can fail.

RCM involves identifying critical assets, determining their required functions, defining the types of failures that can affect these functions, and identifying the failure modes and root causes. Subsequently, a maintenance strategy is chosen using the RCM decision tree, which considers failure consequences, intervals for activities, responsible personnel, and instructions for implementation (Figure 1.4).

For non-critical assets, the preventive maintenance optimization (PMO) method is applied [16]. Implementing PMO requires identifying, categorizing, and consolidating failures based on existing maintenance activities and identifying any failures that lack a maintenance plan. The maintenance plan is then selected using the RCM decision tree, considering failure consequences, and PMO instructions are developed, executed, and trained.

1.3.5 Lifecycle Management Activities

The fifth stage in the maintenance excellence model, focuses on managing lifecycle activities. Lifecycle costs encompass all visible and hidden costs associated with designing, procuring, constructing, maintaining, repairing, refurbishing, and replacing assets [17]. Understanding these costs is critical for ensuring long-term economic viability and sustainability in the management of large-scale infrastructure. For instance, in the context of bridge management, lifecycle cost analysis (LCCA) can be applied to calculate the total cost of ownership over a bridge's operational lifespan, including the costs of periodic inspections, retrofitting, and eventual replacement.

LCCA helps engineers and asset managers determine the optimal time to replace assets, make renovation decisions, estimate the economic lifespan of structures, assess the viability of new technologies, and forecast future budget requirements. It provides a framework for evaluating different maintenance strategies and choosing the one that offers the best balance between cost and performance over the asset's lifespan. By focusing on lifecycle costs, managers can make informed decisions on whether to repair, upgrade, or replace infrastructure components, thereby reducing unnecessary expenses and extending the service life of assets.

A key aspect of LCCA is its role in transitioning from corrective maintenance—where actions are only taken after failure occurs—to predictive and preventative maintenance, which anticipates potential failures before they happen. Corrective maintenance can be costly and disruptive, especially in critical infrastructure like power plants, transportation networks, and water supply systems. In contrast, predictive

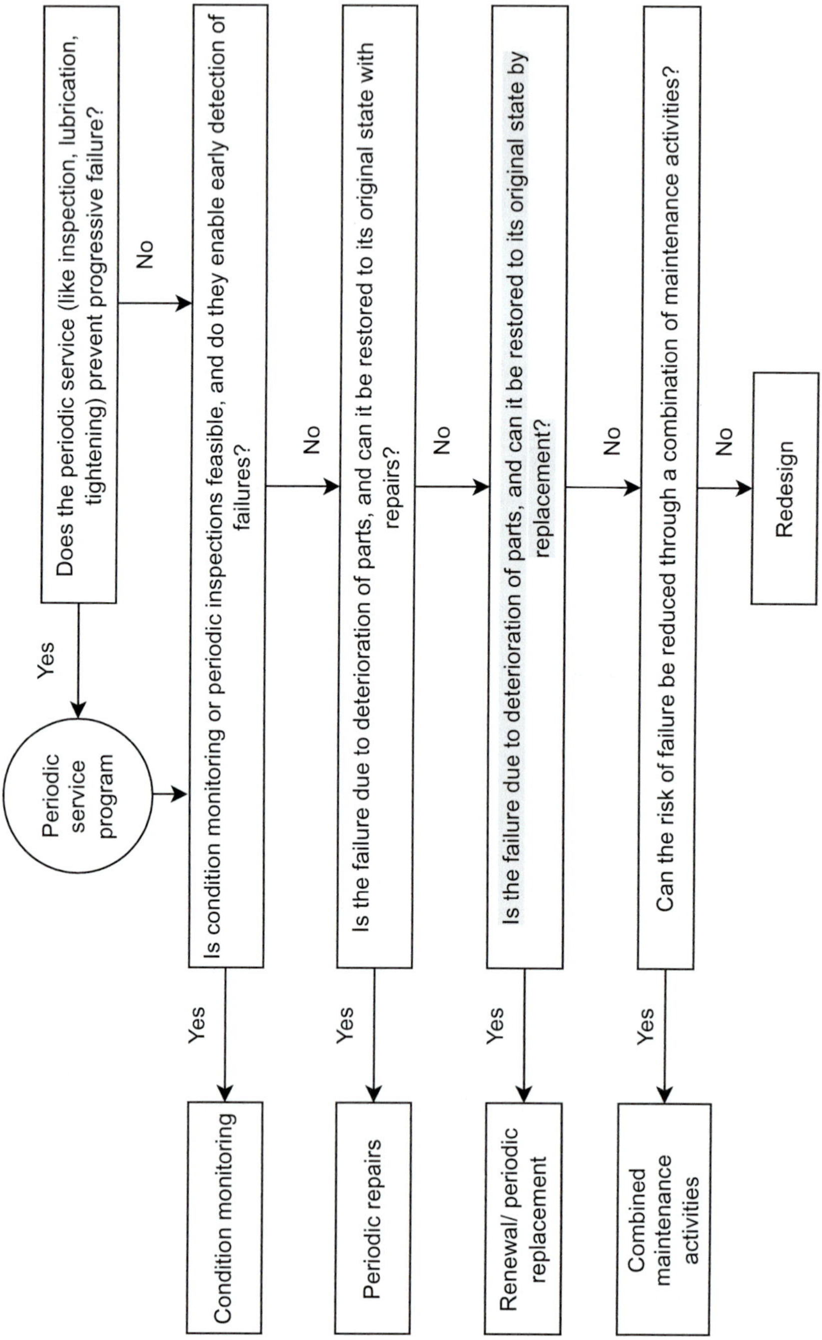

FIGURE 1.4 Example of RCM logic for determining maintenance activities.

maintenance, supported by LCCA, helps in optimizing maintenance schedules and reducing unexpected downtimes.

For example, the U.S. Federal Highway Administration (FHWA) conducted a study using LCCA to determine the most cost-effective strategies for managing highway pavements. They concluded that incorporating LCCA in the early design phases reduced total costs by 10%–15%, while ensuring the longevity and performance of the infrastructure [18].

Failure patterns can vary significantly, depending on the type of asset and its operational environment. For instance, if the probability of failure is high both after manufacturing (due to poor quality control) and toward the end of its lifespan (due to material deterioration), the failure pattern follows a "bathtub curve," where failure rates are high at the beginning and end but relatively constant in between [19]. This is common in mechanical systems and components subjected to fatigue and wear.

The bathtub curve model is helpful in determining maintenance and inspection schedules. Early-life failures, or "infant mortality" failures, are often addressed by more stringent quality assurance during production, while late-life failures can be managed by condition-based monitoring and planned replacements.

However, many failures are not time-dependent and cannot be predicted through traditional lifecycle analysis alone. In complex systems, failures may arise due to unexpected conditions such as environmental factors, design flaws, or operational overloads. These unpredictable failures highlight the limitations of time-based maintenance and the necessity for real-time condition-based monitoring and data analytics to manage critical infrastructure. For predictable failures, expanding the scope of condition-based monitoring allows predictive maintenance to be implemented more effectively.

The application of Artificial Intelligence (AI) and machine learning for lifecycle management has recently gained significant traction, enabling more accurate predictions of asset failure times and thus reducing unexpected failures and costs [20]. By leveraging advanced data analytics, AI can process vast amounts of sensor data and identify patterns that would be impossible to detect manually. This has transformed maintenance practices across industries, shifting from reactive to proactive approaches.

1.3.6 Performance Management and Continuous Improvement

In the sixth stage of the maintenance excellence model, performance measurement becomes a crucial activity. This stage involves the systematic evaluation of the effectiveness of maintenance activities and interventions, aimed at ensuring that the structure continues to meet its operational requirements safely and efficiently.

Performance measurement is not just a one-time assessment but part of an ongoing process. It typically includes key performance indicators (KPIs) such as structural integrity, serviceability, load-bearing capacity, and overall system safety. These KPIs are based on data collected from sensors, visual inspections, and other diagnostic tools employed during earlier SHM stages. By continuously measuring these factors, stakeholders can gauge whether the implemented maintenance strategies are achieving the desired outcomes.

An integral part of this phase is the preparation of evaluation reports. These reports provide a comprehensive overview of the structure's current health, identifying any potential risks, deficiencies, or areas where performance has degraded over time. Additionally, these reports assess the effectiveness of previous interventions and maintenance actions, offering evidence of improvements or identifying where adjustments may be required.

One of the most significant benefits of the performance evaluation stage is its role in facilitating continuous improvement. The data and insights gathered allow maintenance teams to refine their strategies by identifying what has worked and what hasn't. This process supports an adaptive approach, where maintenance methods are continuously fine-tuned to better suit the specific needs of the structure.

Incorporating continuous feedback into the maintenance cycle also supports the adoption of innovative methods. Technological advancements, such as the integration of machine learning algorithms for predictive maintenance, the use of drones for detailed visual inspections, and more advanced sensors that provide real-time data, are all examples of how SHM is evolving. Each iteration of the performance evaluation helps in assessing whether new technologies and methodologies can enhance the reliability and efficiency of monitoring systems.

For instance, a case study on the Humber Bridge in the UK demonstrated how incorporating real-time SHM data allowed engineers to shift from a traditional, time-based maintenance model to a more effective CBM approach. This resulted in a more accurate prediction of maintenance needs, reducing both costs and downtime while improving the overall safety of the structure [21].

By integrating the lessons learned from these evaluations, maintenance strategies evolve in response to changing conditions, ensuring that they remain effective even as the demands on the structure increase or the environment changes. The adoption of innovative methods, driven by continuous performance evaluation, fosters a culture of excellence within the maintenance process. It encourages stakeholders to strive for better safety outcomes, operational efficiency, and cost-effectiveness.

1.4 ROOT CAUSE ANALYSIS

As stated in the third stage of the maintenance excellence model, it is crucial to prevent early failures by addressing defects in critical assets, resulting in rapid performance improvements. Defects that occur during the operational phase should not be considered routine maintenance activities but should be analysed to identify their root causes and prevent recurrence. Early identification of such issues not only mitigates long-term damage but also significantly reduces operational downtime and maintenance costs. For example, early detection of a minor crack in a structural component of a bridge can prevent catastrophic failure, which could lead to severe economic losses and endanger human lives [22].

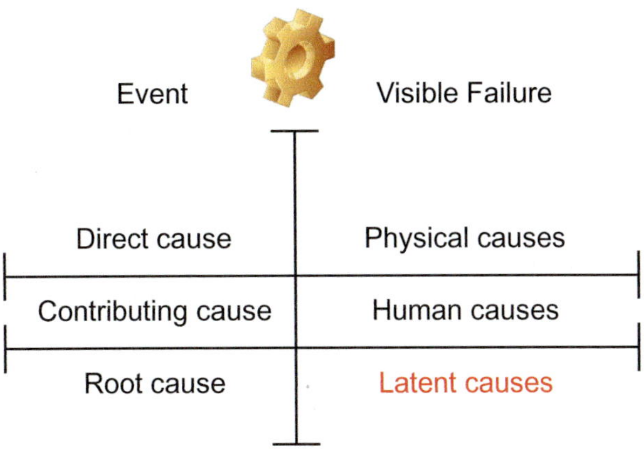

FIGURE 1.5 Root causes of failures.

Root cause analysis (RCA) is a systematic method for identifying the fundamental causes of defects so that appropriate corrective actions can be taken, thereby preventing similar failures in the future. RCA is particularly important in the maintenance excellence model, where the goal is not only to detect damage but also to understand why it occurred, how it can be prevented, and what implications it has for the overall system integrity. The "Why–Why" technique, which involves repeatedly asking "why" until the root cause is identified, is one of the simplest yet most effective tools in RCA [23].

In the maintenance excellence model, RCA is used to identify failures in critical infrastructure such as bridges, dams, and buildings, where small defects, if left unattended, can lead to structural collapse. For example, in the I-35W Mississippi River Bridge collapse in 2007, subsequent RCA revealed that a design flaw, coupled with long-term fatigue and material degradation, was the primary root cause of the disaster.

Robert Nelms, a pioneer in RCA across various industries, classified failure causes into three categories: physical, human, and hidden causes (Figure 1.5). This classification is highly relevant to structural health monitoring, where defects often arise from a combination of these factors.

- *Physical causes*: These include mechanical or material issues such as corrosion, erosion, resonance, imbalance, fatigue, and excessive loading. For example, corrosion of steel reinforcements in concrete bridges, due to chloride penetration from road salts, is a common physical failure cause that can compromise the integrity of the structure [24]. Similarly, excessive vibration in industrial machinery can cause resonance, leading to fatigue cracks in critical components, which can then propagate and result in catastrophic failure if not addressed in time.
- *Human causes*: These involve human error or negligence, such as oversight, mistakes, ignoring defects, accepting issues, inexperience, stubbornness, procrastination, and intentional errors. A notable case study of human error in structural failure is the Hyatt Regency Walkway collapse in Kansas City in

1981 [25]. A design change during construction, without proper engineering oversight, led to a catastrophic failure of the suspended walkways, killing over 100 people. This failure was traced back to human error in the design approval process and a lack of thorough analysis [25].

- *Hidden causes*: These are more insidious and involve underlying systemic issues, such as inadequate training, inappropriate tools, incorrect prioritization, lack of procedures, weak technical specifications, lack of motivation, and poor intra-organizational communication. For instance, the Fukushima Daiichi nuclear disaster in 2011, although primarily triggered by a natural event (tsunami), was exacerbated by hidden causes such as inadequate emergency preparedness, poor communication among stakeholders, and insufficient training of operational staff [26]. In the maintenance excellence model, hidden causes can include inadequate monitoring strategies or improper interpretation of data, which may lead to missed warning signs of impending failures.

The U.S. Department of Energy categorizes failure causes as direct, contributing, or root causes. The "fishbone" or cause-and-effect diagram can help identify these causes [27]. The Reliability Centre of America, a reputable research organization, developed the "PROACT" method, combining the "Why–Why" approach with Nelms' model. This method includes five steps: data collection, forming an analysis team, creating a cause tree (physical, human, hidden), proposing solutions, and tracking results.

1.4.1 Case Study: RCA of the I-35 Bridge Failure

The sudden collapse of bridges, causing casualties and financial losses, occasionally occurs even in developed countries. The I-35 Bridge in the United States was a 581-m continuous truss bridge over the Mississippi River, constructed in 1967 [28]. It collapsed in 2007, resulting in 13 fatalities, 145 injuries, and traffic disruption for 140,000 vehicles per day for 414 days, with an estimated total damage of $460 million (Figure 1.6). Before its failure in 2003, the bridge was classified as structurally deficient but meeting minimum standards with a 50% rating.

The physical cause of the collapse was found to be an inadequate connection plate thickness of 13 mm (Figure 1.7). The contributing factors included the addition of 51 mm of concrete to the deck over time, leading to a 20% increase in dead load, and the placement of construction equipment at the weakest point just before failure. The human cause was the oversimplification of connection plate analysis, which failed to consider its significance in evaluating the bridge's load-bearing capacity.

To address this root cause, a connection plate load evaluation guide was developed to educate bridge assessment engineers [9]. Additionally, the replacement bridge incorporated a smart monitoring system with 323 sensors to detect issues before they became catastrophic. These sensors measured strain, force, vibration, traffic flow, speed, and other factors, providing real-time data for proactive maintenance (Figure 1.8).

FIGURE 1.6 I-35 bridge collapse in 2007, USA, resulting in 13 fatalities [28].

FIGURE 1.7 Inadequate thickness of the connection plate [28].

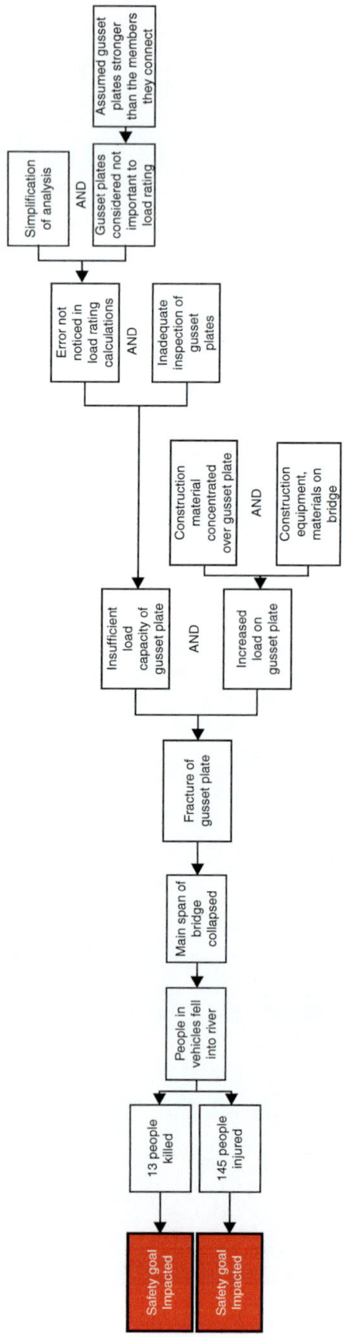

FIGURE 1.8 Root cause analysis map of the I-35 bridge collapse [28].

1.5 SUMMARY

This chapter has established the role of SHM within asset maintenance management. It introduced a simplified six-step excellence model based on ISO 55000: cultural transformation, asset information management, preventing premature deterioration, optimizing maintenance programs, lifecycle management activities, and performance management.

One key focus of this book is to explain how condition monitoring with sensors, particularly for structures such as bridges, plays a pivotal role in implementing CBM. Furthermore, another aim is to demonstrate how condition monitoring aids in RCA for critical defects. By understanding the position of SHM within the asset maintenance management framework, stakeholders can recognize its importance and recommend its appropriate use to asset owners.

QUESTIONS

1. What are the consequences of the degradation of the concrete cover in a reinforced concrete beam?
2. What type of maintenance activity does painting a steel structure fall under?
3. What is the failure pattern of corrosion in a steel member?
4. What could be the root cause of fatigue cracks at the connection between two members of a steel truss?
5. If early fault monitoring leads to early detection of failure, which maintenance strategy should be employed?

REFERENCES

[1] "EN 13306 Maintenance—Maintenance Terminology," 2017.

[2] A. Garg and S. G. Deshmukh, "Maintenance Management: Literature Review and Directions," *Journal of Quality in Maintenance Engineering*, vol. 26, no. 1, pp. 3–24, 2020.

[3] International Organization for Standardization, "ISO 55000:2014 Asset Management—Overview, Principles, and Terminology," 2014.

[4] D. Arler, "Asset Management Implementation on the Sydney Harbour Bridge," *Australian Journal of Civil Engineering*, vol. 16, no. 2, pp. 123–136, 2018.

[5] *Bridge Maintenance Manual: Sydney Harbour Bridge Maintenance Strategy*. Transport for NSW, 2019.

[6] T. Brooks, "Digital Asset Management in Structural Health Monitoring," *Journal of Infrastructure Systems*, vol. 26, no. 4, p. 04020029, 2020.

[7] Transport for NSW, "Sydney Harbour Bridge Annual Maintenance Report," 2021.

[8] R. Smith and B. Hawkins, *Lean Maintenance: Reduce Costs, Improve Quality, and Increase Market Share*. Elsevier, 2018.

[9] J. Moubray, *Reliability-Centered Maintenance*. Industrial Press Inc., 1997.

[10] C. Barton, "The Impact of Maintenance Culture on Asset Management Performance," *Asset Management & Maintenance Journal*, vol. 34, no. 2, pp. 15–27, 2019.

[11] A. Patra, D. Acharya, and S. Bhattacharya, "Maintenance Excellence Model Implementation at Tata Steel," *International Journal of Quality & Reliability Management*, vol. 34, no. 2, pp. 291–308, 2017.

[12] Transport for London (TfL), "London Underground Asset Management Strategy Summary," 2016.
[13] International Organization for Standardization, "ISO 14224:2016—Petroleum, Petrochemical and Natural Gas Industries—Collection and Exchange of Reliability and Maintenance Data for Equipment," 2016.
[14] A. Marquez, P. Moreu, A. Sola, and J. Fernandez, "Criticality Analysis for Maintenance Purposes: A Study for Complex In-service Engineering Assets," *Quality and Reliability Engineering*, vol. 32, no. 2, 2015. DOI:10.1002/qre.1769
[15] Metropolitan Transportation Authority (MTA), "Subway Action Plan Report," 2019.
[16] R. K. Mobley, *Maintenance Fundamentals*, 2nd ed. Elsevier, 2004.
[17] A. Boussabaine and R. Kirkham, *Whole Life-Cycle Costing: Risk and Risk Responses*. Wiley-Blackwell, 2004.
[18] J. Walls and M. R. Smith, "FHWA-SA-98-079: Life-Cycle Cost Analysis in Pavement Design," 1998.
[19] C. E. Ebeling, *An Introduction to Reliability and Maintainability Engineering*, 3rd ed. Waveland Press, 2019.
[20] J. Mattioli, P. Perico, and P. O. Robic, "Artificial Intelligence Based Asset Management," in *IEEE 15th International Conference of System of Systems Engineering (SoSE)*, 2020.
[21] J. Brownjohn, "Structural Health Monitoring of Civil Infrastructure," *Philosophical Transactions A: Mathematical, Physical and Engineering Sciences*, 2016. DOI: 10.1098/rsta.2006.1925
[22] H. Sohn, C. R. Farrar, F. M. Hemez, and J. J. Czarnecki, "A Review of Structural Health Review of Structural Health Monitoring Literature 1996-2001," in *Third World Conference on Structural Control*, Como, Italy, 2002.
[23] B. Andersen and T. Fagerhaug, *Root Cause Analysis: Simplified Tools and Techniques*. ASQ Quality Press, 2006.
[24] S. E. Hussain and A. S. Al-Gahtani, "Chloride Threshold for Corrosion of Reinforcement in Concrete," *ACI Materials Journals*, vol. 93, no. 6, pp. 534–538, 1996.
[25] R. D. Marshall, E. Pfrang, E. Leyendecker, and K. Woodward, "Investigation of the Kansas City Hyatt Regency Walkways Collapse," 1982. NBSIR 82-2465 (A).
[26] C. Perrow, "Fukushima and the Inevitability of Accidents," *Bulletin of the Atomic Scientists*, vol. 67, pp. 44–52, 2011.
[27] US Department of Energy (DoE), "DOE-NE-STD-1004-92, Root Cause Analysis Guidance Document," 2010.
[28] Minnesota Department of Transportation, "Collapse of I-35W Highway Bridge," 2007.

2 Condition-Based Maintenance of Structures

2.1 INTRODUCTION

As discussed in the previous chapter, implementing a preventive maintenance strategy is crucial for achieving excellence in asset maintenance (Figure 2.1). The ability to detect and understand the presence, location, and severity of structural defects has been greatly enhanced with advancements in digital technology, making preventive maintenance more achievable than ever before. Today, the concept of condition-based maintenance, which is rooted in continuous or periodic monitoring, allows infrastructure managers to take preemptive action before defects become critical.

This shift is akin to the evolution seen in other industries, such as in aerospace and automotive industries, where regular health checks and non-destructive testing (NDT) methods have been adopted widely to ensure safety and longevity. Similar advancements are now finding their place in civil engineering, particularly in the inspection and monitoring of bridges, culverts, tunnels, and other structural assets in the transportation sector.

For example, the Sydney Harbour Bridge in Australia has been employing a sophisticated health monitoring system that combines various NDT techniques and real-time data acquisition to detect potential defects early. The system includes sensors to monitor strain, displacement, temperature, and humidity, enabling engineers to assess the structure's condition continually [2]. This has allowed for the early detection of cracks and other defects, significantly extending the bridge's service life and reducing maintenance costs by facilitating timely interventions.

Bridges serve as one of the main assets in roadways and railway lines. They are necessary for crossing natural obstacles or intersections with other routes. Generally, bridges can be categorized into culverts and conventional bridges. If the total length of a bridge is less than 6 m, it is classified as a culvert. The design of culverts is primarily based on hydraulic considerations, aiming to direct floodwaters from one side of an embankment to the other. In culverts, permanent loads are usually greater than live traffic loads. Due to the large number of culverts in a given route, standardized design drawings are available. Engineers select the appropriate culvert design after conducting hydrological and hydraulic studies.

Culverts can be either rigid or flexible from a structural standpoint. In rigid culverts, which are typically made of concrete and constructed in various forms like slab (Figure 2.2), box, arch, frame, or pipe, the culvert itself bears the load. However, in flexible culverts (e.g., corrugated steel sheets), the combination of the culvert and

DOI: 10.1201/9781003494652-2

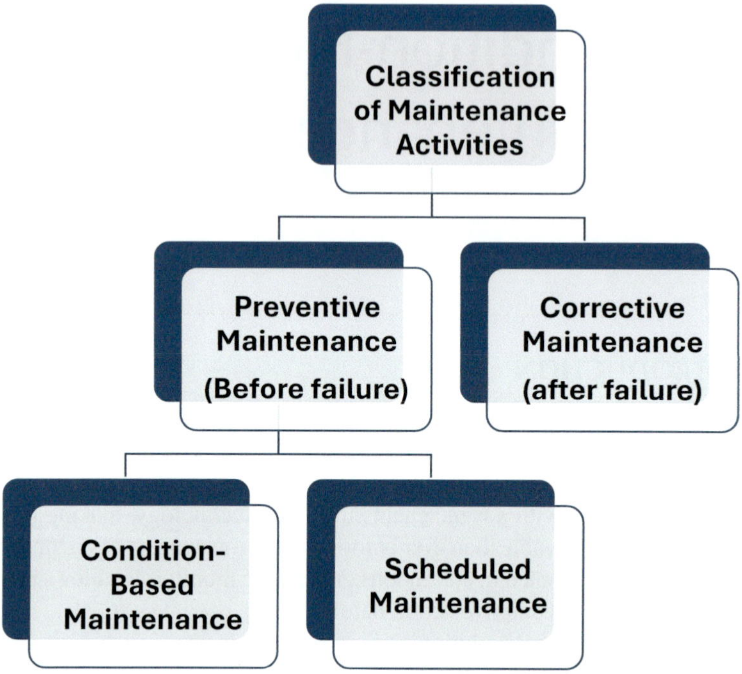

FIGURE 2.1 Classification of maintenance activities according to ISO 14224 [1].

FIGURE 2.2 Buried slab concrete culvert on a railway.

the surrounding soil provides the load-bearing capacity. The interaction between the culvert and the soil is essential in flexible designs, as the soil provides additional support that enables the culvert to carry significant loads.

The primary function of a culvert is its capability to handle floodwaters. During flooding, water often collects at the entrance of the culvert, resulting in slower inflow and faster outflow. This behaviour, known as inlet control, must be managed effectively. In flat areas, water may enter more quickly and exit more slowly, a condition referred to as outlet control. Given the hydraulic behaviour of culverts, it's crucial to monitor both upstream and downstream conditions and perform maintenance activities, such as desilting or erosion control, as needed.

Typically, a corrective maintenance strategy is suitable for culverts, while predictive maintenance strategies are more appropriate for bridges. However, if harsh environmental conditions jeopardize the structural integrity of a culvert, it should be managed similar to a bridge using a preventive maintenance programme.

A study conducted by Lu and Phares [3] provides an example of how preventive maintenance strategies, coupled with SHM systems, can be used effectively for managing culverts and bridges. Their research discusses the integration of SHM systems into a multilayer statewide bridge maintenance and management programme, demonstrating that regular inspections, SHM, and preventive maintenance can significantly improve infrastructure lifespan and safety. This approach proved successful in detecting potential issues early, allowing for timely intervention and reducing long-term maintenance costs.

By adopting a combination of SHM systems, regular inspections, and preventive maintenance strategies, infrastructure managers can extend the life of culverts and bridges while reducing maintenance costs and improving safety.

2.2 BRIDGE CONDITION ASSESSMENT THROUGH INSPECTIONS

To implement preventive maintenance for bridges, inspections of the bridge and its components, as well as the roadway above and below the bridge, are necessary. These inspections provide essential data for assessing the current condition and identifying potential structural issues before they evolve into critical failures. The inspection process, when carried out systematically, ensures the safety, functionality, and longevity of bridge structures. Among the most widely recognized guidelines for this purpose is the *Bridge Inspector's Reference Manual* (BIRM) developed by the Federal Highway Administration (FHWA) of the United States [4]. The BIRM is a cornerstone of bridge inspection protocols and is widely referenced in both the United States and internationally. It serves as a comprehensive guide for conducting inspections and evaluating bridge conditions with a structured approach that ensures consistency and accuracy.

2.2.1 Bridge Inspection Reference Manual

The *Bridge Inspection Reference Manual* [4] provides detailed guidance on inspecting all bridge components, ensuring that inspectors can systematically assess every aspect of the structure. The manual covers a broad range of inspection techniques, evaluation criteria, and documentation practices to ensure uniformity in how bridges are inspected across different regions and by different inspectors.

The manual's primary objective is to ensure the safety of bridges by enabling inspectors to identify deficiencies early on and recommend appropriate repair or maintenance actions. It outlines procedures for conducting both routine and in-depth inspections and provides methodologies for addressing a variety of bridge types, materials, and environmental conditions. Additionally, it aligns with the National Bridge Inspection Standards (NBIS), which set the regulatory framework for bridge inspections in the United States.

The BIRM is divided into multiple sections, each addressing specific aspects of bridge inspection. Key components include:

1. *Bridge types and materials*: The manual covers different bridge designs (such as suspension, arch, girder, and truss bridges) and various construction materials (concrete, steel, timber, masonry). Each bridge type and material presents unique challenges and deterioration mechanisms, and the manual provides tailored inspection guidelines for each. For example, steel bridges are prone to corrosion and fatigue cracking, while concrete bridges often suffer from spalling and chloride-induced corrosion of reinforcing steel.
2. *Inspection frequency and classification*: The BIRM defines inspection intervals based on bridge type, material, traffic volume, and condition. Typically, bridges undergo a routine inspection every 24 months, though some structures in poor condition or those subjected to severe environmental conditions may require more frequent inspections. The manual classifies inspections into routine, in-depth, fracture-critical member (FCM) inspections, and special inspections following events such as natural disasters or vehicle impacts.
3. *Deficiency identification and rating system*: The BIRM provides a systematic approach to rating bridge conditions. Using a numerical system, inspectors assign condition ratings to different bridge components, ranging from 0 (failed condition) to 9 (excellent condition). The manual offers detailed criteria for each rating level, ensuring consistency in assessments. For instance, a steel girder with moderate corrosion and section loss might be rated a 5 (fair condition), while severe rusting and pitting would warrant a lower rating, prompting repairs or further investigation.
4. *Inspection techniques*: The manual categorizes inspection methods into visual and NDT techniques. Visual inspections are the primary method and involve a close-up evaluation of all accessible components. NDT methods, such as ultrasonic testing and magnetic particle testing, are recommended when internal or hidden defects are suspected. The BIRM emphasizes the importance of using advanced tools to complement visual inspections, particularly for critical components that may be subject to fatigue or corrosion.
5. *Inspector qualifications and training*: The BIRM sets forth minimum qualifications for bridge inspectors, stressing the importance of specialized training and certification. It requires inspectors to complete

the FHWA-approved *National Bridge Inspection Training* programme and encourages continuing education in areas such as NDT and structural analysis. This ensures that inspectors stay up-to-date with the latest inspection techniques and safety protocols.

6. *Documentation and reporting*: The BIRM provides standardized formats for reporting inspection findings. These include detailed condition assessments, photographs, sketches, and load rating calculations. Proper documentation is critical for tracking the deterioration of bridge components over time and for making informed decisions about maintenance and rehabilitation. The BIRM also includes guidance on how to store and retrieve inspection records using modern asset management systems, such as the AASHTO Bridge Management System (BMS).

The BIRM outlines a systematic process for conducting a bridge inspection, which typically follows these steps:

1. *Preparation and review of previous reports*: Before starting an inspection, inspectors are required to review previous inspection reports, repair histories, and design drawings to gain a thorough understanding of the bridge's condition and past performance. This helps in identifying recurring issues and components that require particular attention.
2. *Initial walk-through*: The first step in the physical inspection is a general walk-through of the bridge to get an overall sense of its condition. Inspectors assess any obvious signs of distress, such as visible cracks, misalignment, spalling, rust, or other indicators of deterioration.
3. *Detailed component evaluation*: Following the walk-through, the inspector performs a detailed evaluation of each bridge component (deck, superstructure, substructure, bearings, etc.). The BIRM provides checklists and guidelines for evaluating each element. For example, the manual suggests checking for cracks in concrete decks, rust or section loss in steel components, and signs of scour at substructure elements like piers and abutments.
4. *Non-destructive testing* (NDT): If potential problems are identified during the visual inspection, the BIRM recommends performing NDT techniques. For example, ultrasonic testing can be used to detect internal cracks in steel girders, and ground-penetrating radar (GPR) can help identify voids in the substructure or delamination in concrete decks.
5. *Condition rating and load rating*: Once the inspection is complete, inspectors must assign condition ratings to each component based on the BIRM's rating system. The load rating is recalculated if the inspection reveals any deterioration that could impact the bridge's load-carrying capacity. The BIRM provides guidelines for calculating the remaining service life of components based on their condition ratings.
6. *Recommendations for maintenance and repairs*: The final step involves providing recommendations for maintenance, repairs, or replacement. The BIRM

categorizes maintenance actions into urgent (immediate attention required), short term (within one to two years), and long term (within five years) based on the severity of the deterioration and the bridge's importance in the transportation network.

7. *Reporting and recordkeeping*: The inspection findings are documented using the reporting formats provided by the BIRM. This information is submitted to the relevant authorities for review, and the inspection report becomes part of the bridge's historical record for future reference.

The *Bridge Inspector's Reference Manual* [4] is an invaluable resource for bridge inspectors, providing a structured approach for assessing the condition of bridge structures. By following its guidelines, inspectors can ensure that potential issues are identified and addressed before they lead to critical failures. The manual's combination of visual, tactile, and advanced inspection techniques ensures comprehensive coverage of both surface-level and hidden defects, thereby safeguarding the safety and longevity of bridges.

2.2.2 Components of a Bridge

A bridge typically comprises three primary components:

- *Deck*: The part of the bridge that carries traffic, including expansion joints, slabs, sidewalks, railings, drainage systems, and signs. The deck is often subjected to harsh conditions, such as traffic loads, de-icing salts, and temperature fluctuations, making it a priority in inspections. For example, the deck of the San Francisco-Oakland Bay Bridge faced severe corrosion problems due to saltwater exposure, prompting extensive retrofitting efforts [5].
- *Superstructure*: The load-bearing system transfers loads to the supports and includes the main load-bearing elements (girders, cross-beams, longitudinal beams, and connections), secondary members (cross-bracing), and bearings (such as neoprene). The condition of these components is crucial since any defects can directly impact the bridge's load-carrying capacity. A well-documented example is the collapse of the Silver Bridge over the Ohio River in 1967. The failure of an eye bar in the suspension chain, exacerbated by stress corrosion and fatigue, demonstrated the catastrophic consequences of overlooking superstructure integrity [6].
- *Substructure*: The system transferring loads from the superstructure to the ground includes abutments, piers, foundations, and piles. Since these elements are often in contact with soil and water, they are more prone to issues such as scour, corrosion, or settlement. The failure of the Schoharie Creek Bridge in New York (1987) serves as a classic case of substructure failure due to scour, where heavy rainfall caused the undermining of the pier foundation, ultimately leading to the bridge's collapse [7].

2.2.3 Inspection Methods and Tools

Inspections can vary in depth and method, ranging from visual checks to tactile inspections, using simple tools (hammers, probes, measuring tapes, calipers, plumb bobs) or advanced instruments (ultrasonic devices, eddy current testing, acoustic emission testing, infrared thermography, ground-penetrating radar, strain gauges, displacement transducers, accelerometers, etc.). Each tool has its advantages depending on the bridge material and the specific type of damage being investigated.

Visual inspections are the most common and involve assessing the bridge for any obvious signs of damage, such as cracks, rust, or misalignments. While efficient, visual inspections have limitations in detecting sub-surface or internal defects. According to research by Phares et al. [8], visual inspections can fail to identify all critical internal defects.

Tactile inspections allow inspectors to physically feel for defects such as spalling or delaminations, which might not be apparent visually. For instance, in concrete bridges, tactile inspections combined with hammer-sounding techniques are commonly used to detect delaminations in the deck.

2.3 ADVANCED INSPECTION TOOLS

Advanced inspection tools can be categorized into those for local behaviour (focusing on specific defects) and those for assessing the overall behaviour of bridge components. These tools enhance the inspection process, allowing for more accurate and timely assessments of a bridge's structural integrity. Below are some key advanced inspection methods commonly used in bridge inspections:

- *Ultrasonic testing* (UT): Used for detecting internal flaws in steel components. This method utilizes high-frequency ultrasound waves to identify anomalies in materials. For instance, ultrasonic testing was successfully applied to detect cracks in the steel components of the Sydney Harbour Bridge during its regular maintenance programme [9].
- *Infrared thermography* (IRT): Helps identify delaminations or moisture ingress in concrete decks by detecting temperature variations on the surface. IRT can reveal hidden problems by measuring surface temperatures, which can indicate moisture trapped beneath the surface.
- *Ground-penetrating radar* (GPR): Assists in identifying voids or deteriorations within the concrete deck or substructure. GPR is an effective non-destructive testing method that uses radar pulses to image the sub-surface.
- *Acoustic emission testing* (AET): AET involves listening for high-frequency sound waves generated by material deformation or crack propagation. This technique can detect changes in bridge components in real time, allowing for ongoing monitoring. For example, Pullin et al. [10] showed that AE can successfully detect damage in bridge structures.
- *Eddy current testing*: This technique is used primarily for the detection of surface and near-surface defects in conductive materials. By inducing electrical

currents in a material, inspectors can identify cracks and corrosion. Eddy current testing has been utilized effectively on bridges to monitor fatigue cracks in steel girders [11].

- *Laser scanning*: 3D laser scanning technology creates detailed three-dimensional models of bridge structures. This method captures a high-resolution point cloud of the bridge, which can be analysed for structural integrity, deformation, and changes over time. The method was used by Blaskow and Maas to investigate the use of personal laser scanning (PLS) for surveying a concrete bridge structure [12].
- *Fibre optic sensors*: These sensors use light to measure strain, temperature, and other parameters within the structural components of a bridge. Fibre optic sensors can be embedded in concrete or attached to structural elements to monitor real-time changes. A notable example is a structural health monitoring system developed using fibre optic sensors based on fibre Bragg gratings (FBGs) for steel girder highway bridges [13]. The measured strain data from fibre optic sensors were reasonably close to the data measured using traditional strain transducer sensors.
- *Drones and unmanned aerial vehicles* (UAVs): Drones equipped with high-resolution cameras and other sensors have become increasingly popular for inspecting hard-to-reach areas of bridges. They can capture aerial imagery and provide a comprehensive view of the structure without the need for extensive scaffolding. A study by VicRoads in Australia conducted trial inspections using UAVs on four bridges in the North-Eastern Region [14]. The trial found that UAVs provided a safe, compliant, and cost-effective method for visual inspections.

Each of these advanced inspection tools contributes significantly to the safety and maintenance of bridge structures by providing insights that traditional inspection methods may overlook. Implementing a combination of these technologies allows engineers to make more informed decisions about maintenance and repair, ultimately extending the service life of critical infrastructure.

If detailed information about potential defects is unavailable, general behaviour inspection tools (such as accelerometers and displacement sensors) are utilized to identify faulty areas. The I-35W Mississippi River Bridge collapse [15] in Minneapolis, Minnesota, in 2007, highlighted the importance of regular inspections and advanced monitoring techniques. Investigations revealed that the bridge failed due to a combination of undersized gusset plates and increased loads over time, which went unnoticed in routine inspections. This incident led to widespread changes in bridge inspection practices, emphasizing the use of more sophisticated inspection tools to assess structural integrity [15].

2.4 LOAD RATING AND STRUCTURAL HEALTH MONITORING

If any suspicious conditions are observed during a visual inspection, further detailed inspections should be conducted using tactile methods and both simple and advanced tools. The impact of these conditions on load-bearing capacity should be estimated, and load rating calculations performed to determine the remaining capacity of the component. Appropriate maintenance actions can then be determined accordingly.

Crack width monitoring is essential in concrete bridges, as it provides insight into the potential for reinforcement corrosion. Cracks wider than 0.3 mm, exposed to aggressive environments, are particularly concerning and require intervention [16]. Crack monitoring systems, such as strain gauges or fibre optic sensors, are frequently deployed in high-risk areas to detect widening cracks and warn of impending failures.

Access to bridge components is often facilitated by ladders, bridge inspection vehicles (Figure 2.3), scaffolding, rope access equipment (Figure 2.4), or drones. Drones, in particular, have become increasingly popular in recent years due to their ability to inspect hard-to-reach areas without the need for extensive scaffolding or lane closures [14].

Critical elements, areas prone to rainfall accumulation, support regions, and areas subject to bending moments and shear forces should be inspected specifically. These specific areas are essential for identifying potential failure points that can compromise the structural integrity of the bridge. For example, the areas near expansion joints are particularly vulnerable to water infiltration, which can lead to accelerated deterioration of both the deck and superstructure. This highlights the importance of inspecting these locations to prevent moisture-related damage, such as rust formation or freeze–thaw cycles that can exacerbate structural issues.

FIGURE 2.3 Accessing arch bridge components using a bridge inspection vehicle.

FIGURE 2.4 Accessing arch bridge components for displacement sensor installation by rope access.

TABLE 2.1
Decision levels based on the condition of steel members [4]

Condition	Deformation, cracking, corrosion
Non-critical	• Steel members require repainting • Members and connections are adequate. Minor corrosion on the surface is visible.
Critical	• Corrosion and section loss have advanced and are easily visible. • Beam is deformed due to impact. • Less than 25% of bolts in a connection are lost.
Emergency	• Cracks are visible in steel members and connections • Members' deterioration is such that it puts the overall integrity of the structure in jeopardy. • There is a possibility of loss of life for bridge operators. • Deck members have a visible rough surface, excessive deformation, and uneven displacement. • More than 25% of bolts in a connection are lost.

A condition rating system can be applied, typically ranging from good (4), average (3), poor (2), to severe (1), as shown in Table 2.1. This system enables inspectors to quantify the condition of various bridge components, facilitating easier comparison and prioritization of maintenance efforts. Such rating systems are widely used; for

TABLE 2.2
Decision levels based on the condition of concrete members [4]

Condition	Deformation, cracking, corrosion
Non-critical	• Slight and superficial wear is observed in the body. • There may be small cracks and pores on the surface. Cracks are not deep into the concrete members and reinforcements are not exposed. • Hairline cracks with a width of less than 0.1 mm may be observed.
Critical	• Significant and widespread wear is observed in the body. • Cracks are deeper and clearly visible in the structural member. • Reinforcement bars are exposed and corroded, with some deformation or damage. • The member may have clear deformation signs (such as bending or bulging); severe stresses or excessive deflection may be evident.
Emergency	• Severe scouring is present (piles are exposed). • Severe loss of sections and reinforcement, which puts the operation of the structure at high risk. • Substructure members have uneven displacement.

instance, *Bridge Inspector's Reference Manual* uses a 0–9 point scale for rating bridge components [4].

The impact of each component's condition on the load-bearing capacity is assessed, leading to the final load rating of both the superstructure and substructure, as discussed in Chapter 4. This assessment often includes rigorous analysis of historical load data, structural models, and load testing.

Evaluations that consider load-bearing capacity, durability, clearance, bridge width, and scouring risks ultimately guide maintenance decisions (Tables 2.1 and 2.2). These decisions may involve regular inspections, imposing load or speed restrictions, or scheduling special repairs.

Upon the completion of the initial inspection and load rating of the bridge, following the handover from the contractor to the bridge maintenance team, a comprehensive bridge record file should be created. This record serves as the foundational document for the bridge's current structural status. It is essential that this first inspection be accompanied by thorough documentation, including detailed load rating evaluations, to establish a baseline for the bridge's structural condition. Key information such as the bridge's geographic location, age, type, construction materials, and geometric details of its components should be explicitly included in the inspection report. This documentation ensures continuity in maintenance efforts and serves as a reference point for future evaluations.

Given the significance of the bridge in its transportation network, scheduled inspections are planned to identify any changes from the initial condition and to ensure that it remains safe for use. Typically, regular inspections are conducted every 24 months.

Certain bridge components require higher-priority inspections, especially those whose failure could result in the collapse of all or part of the structure. These elements

are classified as fracture-critical members and must be inspected with greater attention to detail. Fracture-critical members are components whose failure would lead to a catastrophic collapse, making it imperative that these parts receive close scrutiny. The inspection should be conducted at close range, ensuring that the inspector can physically touch the components, and if necessary, complemented with NDT methods. NDT allows for the detection of internal flaws without compromising the integrity of the materials. Techniques such as acoustic emission testing or magnetic particle inspection can reveal hidden issues that visual inspections may miss.
The tools required for a comprehensive inspection may include:

- Cleaning tools: Degreasing sprays, brooms, wire brushes
- Inspection tools: Hammers, magnifying glasses, measuring tapes, calipers, crack width rulers, ultrasonic thickness gauges
- Additional equipment: Flashlights, mirrors, spirit levels, thermometers, inspection checklists, pencils, chalk, paint, cameras, and safety boots

As a practical example, the inspection process of a truss bridge is discussed below. This case provides insights into the comprehensive inspection practices required for maintaining such structures effectively.

2.5 CASE STUDY: VISUAL INSPECTION OF NEKA TRUSS BRIDGE

The truss bridge under examination shown in Figure 2.5 spans the Neka River in northern Iran. Constructed in 1999 to replace the previous bridge damaged by flooding,

FIGURE 2.5 Railway warren truss bridge (through truss).

this bridge has a 36-m span. The truss structure is made of ST52 steel and includes 12 sections, each 3 m in length, with a height of 3.3 m and a width of 5.5 m, supporting a single railway track. As a crucial element of the railway network, regular inspections are essential to ensure the safety and longevity of this bridge, particularly given its exposure to environmental elements such as humidity and temperature fluctuations, which can accelerate deterioration.

The inspection of the Neka bridge covered four key sections:

- *Deck*: Including the surface of the bridge, railway track, pedestrian walkway, and other components.
- *Superstructure*: Comprising the upper and lower chords, diagonal members, cross-beams, longitudinal beams, cross-bracing, and elastic bearings.
- *Substructure*: Encompassing the abutments, foundations, and piles.
- *Area Below the Bridge*: Including the Neka River and scouring protection structures. This is not a bridge component but can impact the bridge due to passing traffic (which can crash into the bridge), and running water (potential for scouring).

2.5.1 Deck Inspection

The deck elements, such as rails, timber sleepers, fasteners, and walkways, were examined. Any misalignment or discontinuity in the rail could cause dynamic impact loads and premature deterioration of the superstructure. Such issues can be caused by thermal expansion, differential settlement, or wear and tear from repeated train loads.

During this inspection, special attention was given to the condition of timber sleepers, as they tend to decay or become loose over time, especially in humid environments. The visual inspection rated the deck condition as good, indicating that no immediate action was required.

2.5.2 Superstructure Inspection

The superstructure of a steel truss bridge is critical in ensuring the overall stability of the structure. It includes several key elements:

- *Top and bottom chords*: These form the primary load-carrying members, with the top chord typically in compression and the bottom chord in tension. Their structural integrity is paramount, as any degradation could severely affect the bridge's ability to support loads.
- *Diagonal members and cross-beams*: These distribute loads across the truss and provide lateral stability. Diagonals primarily resist shear forces, while cross-beams support vertical loads from the bridge deck and transfer them to the vertical and diagonal truss members.
- *Longitudinal beams*: Running parallel to the direction of traffic, these beams carry loads from the railway track and distribute them to the cross beams.

- *Cross-bracing*: Provides torsional rigidity to the structure and prevents the bridge from twisting under asymmetrical loads or wind forces.
- *Elastic bearings*: Located between the superstructure and substructure, these components accommodate movements caused by thermal expansion, shrinkage, or live load effects, reducing stress on the superstructure.

Each of these components plays an integral role in maintaining the overall load path from the deck to the substructure, and any failure within this system could compromise the bridge's structural integrity. Regular inspection and monitoring are necessary to ensure that defects do not develop into critical issues.

Steel truss bridges, despite their robust design, are vulnerable to several recurring defects, particularly under harsh environmental conditions or heavy usage. The following issues are often observed:

- Corrosion

Corrosion is the most pervasive problem in steel bridges, particularly in regions with high humidity or proximity to saltwater environments. The presence of moisture accelerates the oxidation of steel, leading to rust formation, which weakens the cross-sectional area of structural members. In coastal areas, airborne salts can exacerbate this issue by penetrating into protective coatings, leading to localized pitting corrosion. The collapse of the Silver Bridge in 1967, which led to 46 fatalities, was attributed to a failure in the suspension chain due to undetected corrosion-induced cracking [6]. Modern approaches to mitigating corrosion include the use of corrosion-resistant materials, such as weathering steel, and advanced protective coatings like epoxy-based systems.

- Fatigue Cracks

Fatigue is a cumulative damage process caused by repeated stress cycles. In bridges, this can result from fluctuating loads due to traffic. Fatigue cracks typically originate at points of stress concentration, such as welds or bolted connections, and can grow over time, ultimately leading to catastrophic failure. A notable example of fatigue-induced failure is the collapse of the Mianus River Bridge in 1983, where undetected fatigue cracks led to the fracture of a pin-and-hanger assembly, causing a section of the bridge to collapse [17]. The integration of SHM systems, such as strain gauges and acoustic emission sensors, can help detect early signs of fatigue cracking, allowing for timely intervention [17].

- Overloading

Overloading, often due to vehicles exceeding legal weight limits, can lead to both short-term and long-term damage. In the short term, excessive loads can deform or buckle steel members, while in the long term, repeated overloading can accelerate fatigue damage. This was observed in the collapse of the I-35W Mississippi River Bridge in 2007, where insufficient load-bearing capacity was a contributing factor [15]. Advanced load monitoring systems, integrated with weigh-in-motion technology, can provide real-time data on vehicle weights, helping authorities to prevent overloading and extend the service life of bridges [15].

- Traffic and Impact Damage

Impact damage from vehicular collisions, particularly on road-under-railway bridges, is another significant concern. Heavy vehicles colliding with bridge elements can cause bending, cracking, or outright failure of steel members. For example, the load on train 2MP9 struck a timber pylon of an over-rail bridge near Great Western, Victoria. The use of impact detection sensors and vibration monitoring can help detect and assess the extent of such damage immediately after an incident.

- Thermal Damage

Steel bridges are particularly susceptible to thermal expansion and contraction due to their high thermal conductivity. In regions with extreme temperature variations, this can lead to stress-induced cracking, especially at connections or areas of restraint. Infrared thermography is increasingly used as a non-destructive testing method to detect thermal anomalies in bridge structures [18].

- Loss of Protective Coatings

Protective coatings, such as paints or galvanizing, play a crucial role in preventing corrosion. Over time, however, these coatings degrade due to exposure to UV radiation, chemicals, or mechanical wear. Once the protective barrier is compromised, steel is exposed to the environment, accelerating the corrosion process. Periodic maintenance of protective coatings is essential to ensure long-term durability. For example, Transport for NSW maintains regular repainting of Sydney Harbour Bridge, being an essential part of ongoing maintenance.

Inspection defects in steel bridges can be identified through visual inspections, which are the primary method of routine monitoring. Inspectors use a variety of tools for close-up tactile and visual examinations, including wire brushes to remove surface rust, magnifying lenses for identifying small cracks, and ultrasonic or dye-penetrant testing for non-destructive inspection of cracks and hidden defects. Advanced technologies, such as LiDAR scanning and drones, have also emerged as valuable tools for capturing 3D models and detecting subtle deformations that might be invisible to the naked eye.

Steel truss bridges are classified as "*fracture-critical structures*" because they contain elements that, if failed, could compromise the entire structure due to lack of redundancy in load paths. These critical components, which include the chords (upper and lower), vertical members, and diagonal elements under tension, bear significant loads without alternative pathways for load distribution. This absence of redundancy makes these components highly susceptible to failure when exposed to defects or damage. The failure of even a single critical member could lead to a catastrophic collapse of the bridge.

To prevent such failures, engineers must conduct regular and thorough inspections, focusing on potential defects that may compromise the structural integrity of fracture-critical components. The following are key areas of concern:

- *Corrosion*: Corrosion can be particularly harmful when it leads to advanced cross-sectional loss of critical members. In extreme cases, the load-bearing capacity of the member may be severely reduced, which can escalate into a critical failure. Coastal or humid environments accelerate this process, as seen in the Silver Bridge collapse in 1967 due to undetected corrosion [6]. Therefore, identifying corrosion early is essential, especially in steel truss bridges exposed to corrosive agents like road salts or seawater.
- *Visible cracks*: Fatigue cracks, especially around connections or welds, are often precursors to catastrophic failure. These cracks tend to initiate at stress concentration points and can propagate over time if undetected. Near collapse of Hernando DeSoto bridge in Arkansas is a great example [19]. While this bridge didn't fully collapse, it experienced a severe crack that could have led to catastrophic failure, highlighting the importance of early crack detection [19].
- *Beam deformation*: Twisting often results from overloading or impact, such as a vehicular collision with bridge elements or even misalignment during initial construction. Such deformations compromise the geometry of the bridge, leading to uneven stress distributions. For example, the collapse of the Quebec Bridge in 1907 was partly due to undetected deformations that led to structural failure [20].
- *Missing bolts*: Loose or missing fasteners are a significant concern, as they can indicate underlying issues with the connections between structural elements. Missing bolts can lead to joint failures, reducing the structural continuity necessary for load transfer. The consequences of this can range from localized failures to full structural collapse.
- *Potential falling elements*: Loose or unsecured parts, such as connection plates or decorative elements, pose a severe safety risk to vehicles and pedestrians below. In some cases, falling debris can indicate a more serious underlying problem, such as joint instability or member failure.

After identifying defects, inspectors evaluate the condition of each component, as outlined in Table 2.1. Factors such as remaining cross-sectional area, crack length, and deformation severity are considered when evaluating load-bearing capacity. Maintenance decisions are then guided by this assessment, as indicated in Table 2.2.

For minor defects, routine maintenance activities such as applying protective coatings, replacing damaged fasteners, or cleaning rusted areas can be sufficient to mitigate the risk of further deterioration. For severe defects, such as extensive corrosion, particularly in load-bearing members like the lower chords, or fatigue cracks in high-stress areas, more significant interventions are required. These may include imposing load restrictions or even bridge closures until the affected components can be repaired or replaced. In extreme cases, this could mean replacing entire sections of the bridge. A notable example is the emergency closure of the Sherman Minton Bridge in 2011 due to the discovery of severe cracking in critical load-bearing members, which required significant repairs before reopening [21].

In truss bridges, cross-beams are particularly critical if they have flexible connections, if the spacing between them exceeds 4 m, if there are no longitudinal beams in the deck, or if the longitudinal beams are simply supported. These beams are classified as "critical failure components" since they are under bending and their failure could lead to the collapse of the entire bridge or a significant portion of it.

The recent visual inspection conducted on the Neka truss bridge indicated that the condition of the superstructure was rated as poor, as shown in Figures 2.6 to 2.10. This finding highlights the urgent need for remedial maintenance actions.

FIGURE 2.6 Upper chord of the northern truss.

FIGURE 2.7 Connection between upper chord and diagonal members.

FIGURE 2.8 Connection between lower chord and diagonal members.

FIGURE 2.9 Underside view of the Neka Truss Bridge Deck.

2.5.3 Substructure Inspection

The bridge's abutments (Figures 2.11 and 2.12), which are reinforced concrete components, serve the dual purpose of retaining the soil behind them and supporting both the bridge deck and superstructure. Common defects in reinforced concrete abutments include cracks, concrete deterioration, efflorescence, exposed reinforcement bars, and spalling of the concrete. The visual inspection conducted rated the abutment's condition as moderate.

2.5.4 Inspection of the Area Below the Bridge

In this case, the Neka River flows beneath the bridge. The bridge inspection process must verify that sufficient water can pass under the bridge during floods and check for scouring around the foundation of the abutments. The condition of existing hydraulic

FIGURE 2.10 Connection between the lower chord and cross beam of the truss bridge.

FIGURE 2.11 East abutment wall.

FIGURE 2.12 West abutment.

structures, such as the riprap and cut-off walls, should also be inspected. The visual inspection concluded that the condition of the area below the bridge was good.

2.5.5 Evaluation of the Bridge Condition Based on Visual Inspection

The bridge's overall condition was evaluated using a radar chart. Assuming the initial condition of all bridge components was good, the visual inspection conducted after 25 years revealed that material degradation had caused deterioration in some parts of the bridge (Figure 2.13). The deck and the area beneath the bridge were found to be in good condition, while the substructure had declined by one level and the superstructure by two levels. In Chapter 4, the load rating for this case study will be reassessed, considering the changes in the bridge's condition.

2.6 Condition Monitoring Using Advanced Sensory Tools

Visual inspection is the most common and fundamental condition-based maintenance activity. While it's often sufficient for identifying obvious defects, its accuracy can be limited, particularly for internal or hard-to-detect issues. This emphasizes the importance of integrating advanced tools into inspection regimes to improve accuracy and ensure structural safety.

Condition monitoring techniques can be broadly classified into two categories:

- *Global monitoring*: Focuses on capturing the overall structural behaviour, such as acceleration, displacement, or vibration responses under operational loads. It helps in identifying changes in the structure's dynamic characteristics, indicating potential damage. An example is using accelerometers to measure the vibration response of a bridge under regular traffic to assess its global structural health.

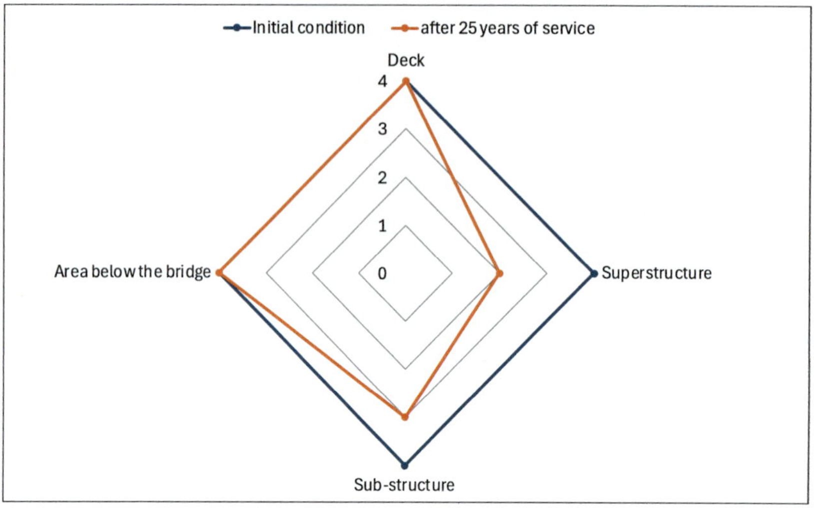

FIGURE 2.13 Condition changes in various bridge components identified through inspection due to material degradation.

- *Local monitoring*: Targets specific defects or damage within a particular component. This approach is suitable when the defect's location is already known or suspected. Techniques like ultrasonic testing and acoustic emission monitoring are examples of local assessments that detect cracks or corrosion in steel members.

A practical case is the Sydney Harbour Bridge, where a combination of global and local monitoring techniques has been employed to detect both overall structural changes and specific defects in critical components [22].

When detailed knowledge about a defect's location is available, local monitoring is appropriate; otherwise, global monitoring is recommended. Monitoring can be either:

- Temporary: Implemented for short-term assessments, such as during load testing or after extreme events (e.g., earthquakes or heavy storms).
- Permanent: Involves installing an SHM system that continuously gathers data to provide real-time insights into the structure's condition over time.

An SHM system is particularly valuable for critical structures, such as long-span bridges, tunnels, or high-rise buildings, where early detection of issues can prevent catastrophic failures. Such systems are especially critical in cases where the consequences of structural failure can lead to significant economic losses, human casualties, or environmental damage. Implementing a permanent SHM system involves a systematic seven-step approach:

1. *Define monitoring objectives*: Establishing clear monitoring objectives is crucial to the success of an SHM system. These objectives vary based on the type of structure, its environment, and the risks associated with its failure. Common objectives include:
   - *Damage detection*: Identifying cracks, corrosion, or fatigue in the structure that could lead to failure.
   - *Load-carrying capacity assessment*: Understanding how much load a structure can safely bear over time.
   - *Effect of environmental conditions*: Monitoring how factors such as temperature, humidity, wind, or seismic activity influence structural performance.
2. *Identify constraints*: Understanding constraints is essential to developing a feasible SHM system. Constraints could include:
   - *Budget*: The cost of installing, operating, and maintaining SHM systems, especially in large structures.
   - *Access difficulties*: Structures such as tall buildings or underwater tunnels may pose significant challenges in terms of sensor installation and maintenance.
   - *Environmental challenges*: In regions prone to extreme weather or seismic activity, the monitoring system itself must be robust enough to withstand these conditions.
   - *Technical constraints*: This involves the resolution and accuracy of available sensors, data storage limitations, and the system's power requirements.
3. *Develop a theoretical model*: A theoretical model predicts how a structure behaves under different loading conditions and is critical for comparing real-time data to expected outcomes.
4. *Design the monitoring system*: Specify input loads (e.g., traffic or wind) and expected output behaviours (e.g., vibration frequencies).
5. *Implement the physical monitoring system*: During the physical implementation phase, appropriate sensors, data acquisition systems, and communication technologies are selected and installed based on the structure's size, location, and critical areas.
6. *Validate the monitoring system*: Validation involves testing the SHM system to verify that it captures accurate and reliable data. This includes performing load tests and analysing sensor data to ensure it matches theoretical predictions. If discrepancies are found, adjustments may be needed in sensor placement or system configuration.
7. *Maintain the monitoring system*: Regular maintenance is critical for the longevity and accuracy of an SHM system. This includes calibrating sensors, updating software, and replacing faulty components. Periodic data review is also necessary to ensure that the system continues to operate effectively.

Figure 2.14 illustrates the various stages of SHM system development, providing a clear roadmap for implementing a comprehensive monitoring system.

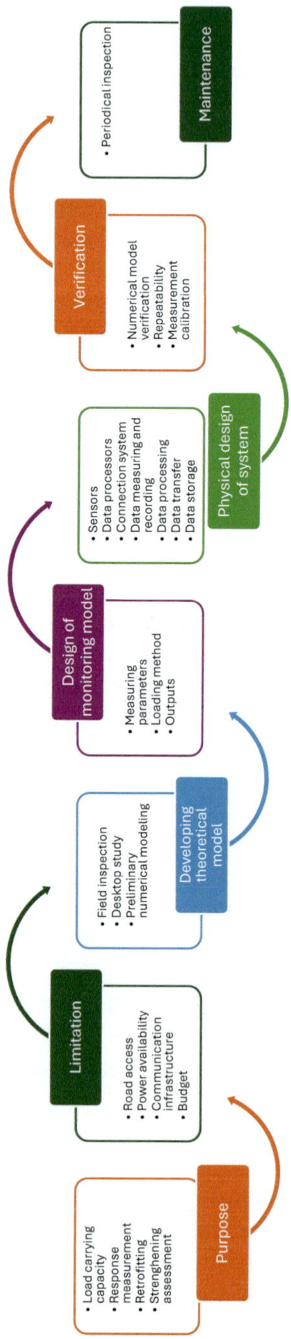

FIGURE 2.14 Stages of the SHM system development.

The SHM system operates within a specific frequency range, and the necessary sensors and technology must be chosen accordingly. Different types of monitoring activities require sensors capable of capturing specific frequency ranges:

- Seismometer and vibrometer operate at frequencies below 20 Hz, making them ideal for ambient monitoring the effects of wind, traffic loads, or seismic activities (Figure 2.15).
- Impact testing of common structures typically ranges up to 200 Hz, often used to assess structural responses to sudden loads, such as drop weight impacts.
- Dynamic testing during the heaviest operational load occurs below 200 Hz, helping monitor how structures behave under maximum load conditions, such as fully loaded trains on a railway bridge.
- Ultrasonic wave transmission and reception operate in the megahertz range, making them suitable for detecting internal flaws, cracks, or corrosion in metallic components.
- Ground-penetrating radar (GPR) operates in the gigahertz range and is widely used to assess sub-surface conditions, detect voids, and evaluate the integrity of concrete elements.

FIGURE 2.15 SHM system for the Ghotoor Bridge under environmental vibrations.

When implementing condition monitoring using advanced sensing technologies, several key questions must be addressed to ensure a comprehensive and effective monitoring campaign:

- Sensor selection
  Which type of sensor should be used (e.g., strain gauges, accelerometers, ultrasonic transducers)? What is its sensitivity, accuracy, and operational frequency range?
- Sensor placement
  Where should sensors be installed to capture the most relevant data? Sensor placement strategies must be tailored to capture critical data points across a structure. For example, strain gauges are typically positioned in areas of maximum stress, such as at the midpoint of beams or near support connections. Accelerometers are often installed at locations of high modal response, where the structure exhibits its largest vibrational response, such as bridge midspans or the top of tall buildings.
- Power supply
  How will the sensors be powered? Providing reliable power to sensors is one of the logistical challenges of long-term SHM. Solar panels are an effective solution in remote areas with abundant sunlight, while batteries are suitable for shorter-term monitoring campaigns.

  Hardwired connections may be required in environments with high electromagnetic interference or when uninterrupted power is critical, such as in nuclear power plants where SHM systems are essential for ensuring operational safety.
- Measurement conditions
  Under what loading conditions should measurements be taken? Consider operational loads (e.g., passing trains or vehicles), environmental conditions (e.g., temperature or wind), and event-based triggers (e.g., earthquakes or storms).
- Sampling frequency
  How many readings should be taken per second, and how much data will be generated? The appropriate sampling frequency depends on the type of load and the structural response. Dynamic events, such as vehicle or train crossings, require high sampling rates (100–200 Hz) to capture transient vibrations accurately. In contrast, monitoring long-term trends, such as thermal expansion or shrinkage, may require much lower sampling frequencies, sometimes as low as 1 Hz.
- Monitoring duration
  How long will the measurement campaign last? Will it be temporary (e.g., post-event assessment), periodic (e.g., monthly checks), or continuous? The length of a monitoring campaign can vary based on the goals. Post-event assessments, such as after an earthquake or extreme weather, might be short term. In contrast, monitoring critical infrastructure like the Tsing Ma Bridge

in Hong Kong, which is continuously monitored, requires long-term solutions [23]. The ability to adapt the monitoring duration to changing structural conditions is an essential feature of SHM systems.

- Sensor calibration
  How will sensors be calibrated to maintain accuracy? Regular calibration is necessary to maintain the accuracy of sensors, particularly for long-term monitoring. For example, the SHM system deployed on the Vasco da Gama Bridge in Portugal involves routine calibration to ensure the accelerometers and strain gauges provide accurate readings over the years [24].
- Data transmission and analysis
  How will data be transmitted to a central repository for analysis? Options include wireless communication (Wi-Fi, Zigbee, or LoRaWAN), fibre optics, or manual data retrieval.
- Health and defect identification
  How will the structure's health, defect location, severity, and remaining life be determined from the collected data? Determining the structural health involves sophisticated algorithms that process the sensor data to detect, locate, and quantify damage. For instance, the SHM system on the Jindo Bridge in South Korea uses artificial neural networks (ANN) to interpret accelerometer and strain gauge data, allowing for the identification of damage at a optimum level [25].
- Validation and verification
  How will the accuracy of the analysis be validated? This might involve comparison with theoretical models, previous inspection data, or load testing.
- Operational and environmental conditions
  How do different operational conditions (e.g., axle loads, speeds) and weather impact the monitoring results? These factors must be considered to avoid false alarms or misinterpretations.
- Data management
  What kind of management dashboards should be developed for data display and storage? Effective data management tools, such as cloud-based dashboards, can visualize trends and detect changes in structural behaviour. A prominent example is the use of the SHM system on the Oresund Bridge between Sweden and Denmark, where data is continuously collected, processed, and visualized on a dashboard that allows engineers to monitor the bridge's health in real-time [26].

The experience gained from monitoring acceleration and strain in road and railway bridges, as well as the monitoring of vertical and lateral forces in railway lines and vibrations from subway trains, provides valuable insights. The skills and knowledge obtained from analysing such data are discussed in the subsequent chapters. The reader is encouraged to consider the questions raised and compare their current understanding with the insights provided at the end of this book.

2.7 SUMMARY

To implement predictive maintenance for bridges, inspections, condition assessments, and load-bearing evaluations must be conducted at regular intervals (e.g., every two years). The simplest form of inspection involves a visual assessment of the bridge and its components, including the structure above and below the bridge deck. Generally, a bridge can be divided into three parts: deck, superstructure, and substructure.

As a case study, the Neka Truss Bridge was inspected in four sections: deck, superstructure, substructure, and the area beneath the bridge. The deck includes the railway track and pedestrian walkway. The superstructure consists of the top and bottom chords, diagonal members, cross-beams, and lateral braces, as well as elastic supports. The substructure encompasses the abutments and foundations, while the area beneath the bridge includes the Neka River and its scour protection structures.

If doubts arise during a visual inspection, a more detailed step-by-step inspection of the suspected components must be conducted. This involves a combination of visual, tactile, and both simple and advanced instrument-based inspections. Following this, the impact of the condition on the load-bearing capacity of the bridge components is estimated, and the remaining capacity of the structure is determined through load rating. Based on these assessments, appropriate maintenance activities can be scheduled.

For a thorough inspection, direct access and cleaning of the inspection area are required. If access to certain components is difficult or impossible, advanced instruments can be used to monitor the overall behaviour of bridge elements, allowing for an indirect assessment of the inaccessible parts.

Advanced monitoring tools are categorized into global and local monitoring systems. In global monitoring, the overall behaviour of a structural member, such as its acceleration or displacement during operation, is observed. In local monitoring, a specific area of a member is evaluated using methods like ultrasonic testing. Permanent condition monitoring, aided by instruments, can be implemented by developing a SHM system through a seven-step process:

1. Define the monitoring objectives.
2. Identify the system's constraints.
3. Develop a theoretical model.
4. Design the monitoring system by determining input loads and output responses.
5. Implement the physical monitoring system based on the design and available technology.
6. Validate the system.
7. Maintain the monitoring system during operation.

In the next chapter, details of advanced structural health monitoring systems will be elaborated.

DISCUSSION QUESTIONS

1. Based on visual inspection, how would you assess the condition of the case study in Section 2.5? How does the condition of the bridge components affect the load-bearing capacity of the structure?
2. What are the critical components in the case study of Section 2.5? What maintenance strategies, guided by reliability-centred maintenance (RCM), could be developed for these components?
3. For which component in the case study can condition monitoring tools be used?

REFERENCES

[1] International Organization for Standardization, "ISO 14224:2016—Petroleum, Petrochemical and Natural Gas Industries—Collection and Exchange of Reliability and Maintenance Data for Equipment," 2016.

[2] M. Makki Alamdari, N. Khoa, T. Rakotoarivelo, H. Kalhori, and J. Li, "Structural Health Monitoring in the Sydney Harbour Bridge Using Spectral Moments," in *8th International Conference on Structural Health Monitoring of Intelligent Infrastructure*, 2017.

[3] L. Ping and B. Phares, "Integration of Structural Health Monitoring (SHM) into Multilayer Statewide Bridge Maintenance and Management Practices—SHM-Facilitated Condition-Based Maintenance (SHM-CBM) Prioritization System," 2018.

[4] *FHWA-NHI-23-024: Bridge Inspector's Reference Manual* (BIRM), 2023. U.S. Department of Transportation.

[5] Toll Bridge Program Oversight Committee, "San Francisco Bay Area Toll Bridge Seismic Retrofit and Regional Measure 1 Programs," 2013.

[6] U.S. Army Corps of Engineers, "Silver Bridge Collapse After Action Report," 1968.

[7] "Collapse of New York Thruway (I-90) Bridge over the Schoharie Creek," 1988. National Transportation Safety Board. NTSB/HAR-88/02.

[8] B. Phares, D. Rolander, B. Graybeal, and G. Washer, "Reliability of Visual Bridge Inspection," *Public Roads*, vol. 64, pp. 22–29, 2001.

[9] Transport for NSW, "Sydney Harbour Bridge Annual Maintenance Report," 2021.

[10] R. Pullin, K. Holford, J. Robert, and M. Eaton, "Acoustic Emission Monitoring of Bridge Structures in the Field and Laboratory," *Journal of Acoustic Emission*, vol. 26, pp. 172–181, 2008.

[11] L. H. Ichinose, Y. Kohno, T. Kitada, and M. Matsumura, "Applications of Eddy Current Test to Fatigue Crack Inspection of Steel Bridges," Graduate School of Engineering, Osaka City University, 2007.

[12] R. Blaskow and H. Maas, "Structural Health Monitoring of Bridges with Personal Laser Scanning: Segment-based Analysis of Systematic Point Cloud Deformations," *ISPRS Annals of the Photogrammetry Remote Sensing and Spatial Information Science*, vol. 10, pp. 9–16, 2024.

[13] R. Lu and J. Judd, "Field-Deployable Fiber Optic Sensor System for Structural Health Monitoring of Steel Girder Highway Bridges," *Infrastructures (Basel)*, vol. 7, no. 2, pp. 16, 2022.

[14] R. Hawken, T. Nguyen, and J. Ivanyi, "Bridge Condition Inspections Using Unmanned Aerial Vehicles—A Trial Project," 2015. Austroads Bridge Conference, 10th, 2017, Melbourne, Victoria, Australia

[15] Minnesota Department of Transportation, "Collapse of I-35W Highway Bridge," 2007. Austroads Bridge Conference, 10th, 2017, Melbourne, Victoria, Australia.
[16] ACI Committee 224, "224.1R-07: Causes, Evaluation, and Repair of Cracks in Concrete Structures," 2007.
[17] "Collapse of a Suspended Span of Interstate Route 95 Highway Bridge over the Mianus River," 1984. National Transportation Safety Board, PB84-916203.
[18] B. N. K. Raja, S. Miramini, C. Duffield, M. Sofi, and L. Zhang, "Infrared Thermography Detection of Delamination in Bottom of Concrete Bridge Decks," *Structural Control and Health Monitoring*, 2021. https://doi.org/10.1002/stc.2886
[19] J. E. A. (WJE) Wiss, "Fracture Investigation of I-40 Hernando DeSoto Bridge," 2021. www.ardot.gov/wp-content/uploads/2021/11/Fracture-Investigation-I-40-MS-Rvr-Bridge.pdf
[20] C. Pearson and N. Delatte, "Collapse of the Quebec Bridge, 1907," *Journal of Performance of Constructed Facilities*, vol. 20, no. 1, pp. 84–91, 2006.
[21] "Emergency Temporary Closure of the I–64 Sherman-Minton Bridge Over the Ohio River Between Indiana and Kentucky," Federal Highway Administration (FHWA), DOT, 2012.
[22] *Bridge Maintenance Manual: Sydney Harbour Bridge Maintenance Strategy*. Transport for NSW, 2019.
[23] Y. Q. Ni, H. W. Xia, and J. M. Ko, "Structural Performance Evaluation of Tsing Ma Bridge Deck Using Long-term Monitoring Data," *Modern Physics Letters B*, vol. 22, no. 11, pp. 875–880, 2008.
[24] A. Cunha, "Identification from the Natural Response of Vasco da Gama Bridge," in 22nd International Modal Analysis Conference, 2004.
[25] S. Cho, S. Jang, H. Jo, and K. Mechitov, "Structural Health Monitoring System of a Cable-Stayed Bridge Using a Dense Array of Scalable Smart Sensor Network," in *Proceedings of SPIE—The International Society for Optical Engineering*, 2010.
[26] B. Peeters, G. Couvreur, O. Razinkov, C. Kündig, H. Van der Auweraer, and G. De Roeck, "Continuous Monitoring of the Øresund Bridge: System and Data Analysis," *Structure and Infrastructure Engineering*, vol. 5, no. 5, pp. 395–405, 2009.

3 Advanced Condition Monitoring Tools

3.1 INTRODUCTION

As covered in Chapter 2, when accessing certain bridge components for visual inspection is difficult, or when visual assessments fail to provide sufficient information, advanced monitoring tools become essential for either direct or indirect evaluation of the structure's condition. Various sensors are available and can be selected based on the requirements of the project. Once the structure is instrumented, the gathered raw data must be processed before it can be used for assessment purposes. This chapter covers various types of sensors available for structural health monitoring and different methods to process raw data, and concludes by providing a case study of load testing of Neka truss bridge.

3.2 GLOBAL VS LOCAL MONITORING TOOLS

Advanced monitoring tools can be broadly categorized into two types: global monitoring tools and local monitoring tools. Global monitoring tools provide a comprehensive assessment of a structure's overall performance. They can capture large-scale defects or anomalies that affect the structural behaviour as a whole. For instance, measuring the mid-span displacement of a bridge girder using displacement sensors provides valuable insights into the overall structural condition. Displacement sensors, which measure shifts in performance even when defects are located far from the sensor, could have provided early warnings of abnormal displacements that signalled deeper structural issues.

Additionally, accelerometers are often used in the context of global monitoring. These devices measure the dynamic response of structures, such as vibration or sway, induced by external forces like wind or seismic activity. For example, Lin et al. focused on cost-effective Micro-Electro-Mechanical Systems (MEMS)

DOI: 10.1201/9781003494652-3

accelerometers for building safety monitoring and conducted comprehensive parameter analysis of sensors to investigate their suitability for rapid after-earthquake assessment [1].

Conversely, local monitoring tools focus on specific, isolated issues, such as cracks in welds or corrosion in certain joints. Tools such as ultrasonic sensors are commonly used for this purpose. Ultrasonic testing (UT) employs high-frequency sound waves to detect internal defects like cracks or voids in materials. For example, this method is extensively used in the maintenance of offshore oil platforms to assess the integrity of welds and pressure vessels. By detecting cracks early on, ultrasonic testing significantly reduces the risk of failures that could lead to catastrophic events. Baldwin et al. highlight the efficacy of UT in detecting fatigue cracks in steel girders subjected to cyclic loading, an application that has proven essential in monitoring fatigue-prone structures such as bridges [2].

Similarly, acoustic emission testing (AET) is another local monitoring technique applied to detect crack propagation in real time. AET is commonly used in nuclear power plants to monitor critical components, such as reactor pressure vessels. For instance, AET has successfully detected crack growth in welds, enabling timely interventions that prevent potential radiation leaks. Pullin et al. [3] have provided extensive research on the use of acoustic emission in monitoring crack growth in both concrete and steel structures, emphasizing its role in early damage detection and prevention of structural failure.

While global monitoring tools, such as displacement sensors and accelerometers, offer a broader perspective by capturing the overall behaviour of a structure, local monitoring tools, such as ultrasonic testing and acoustic emission testing, are crucial for diagnosing specific defects when the damage location is known. This distinction is vital in situations where the location of defects is uncertain, as global monitoring allows for the detection of performance shifts that may indicate underlying issues.

In practical terms, both global and local monitoring methods are often combined to ensure comprehensive structural health monitoring. For example, the Tsing Ma Bridge in Hong Kong utilizes both types of monitoring tools. A long-term structural health monitoring system that includes displacement, strain, and acceleration sensors has been installed to capture the overall performance of the bridge under dynamic loads like typhoons and high traffic volumes. In parallel, local techniques such as ultrasonic testing are used periodically to check critical joints for cracks. This integrated approach ensures that both large-scale structural behaviours and localized defects are identified and addressed, preventing unexpected failures.

In this book, the focus is primarily on global monitoring tools, which offer a broader perspective of the structural condition by measuring parameters that reflect the overall behaviour of the structure (Figure 3.1). While local monitoring tools are effective when the damage location is known, global monitoring techniques are more practical when the location of defects is uncertain and an assessment of structures response is required.

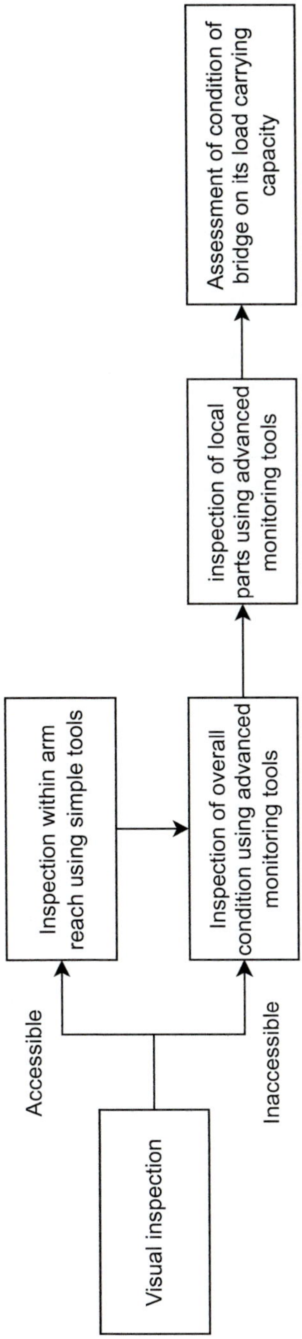

FIGURE 3.1 Role of global monitoring tools in structural health assessment.

3.3 TYPES OF GLOBAL MONITORING SENSORS

Various techniques are available to measure global parameters like strain, displacement, velocity, and acceleration, including:

- *Laser-based sensors*: These are highly accurate, non-contact sensors that measure displacement and can be used for monitoring movements in bridge decks, girders, or towers. For instance, the Zhao et al. [4] study developed a bridge displacement monitoring system based on laser projection-sensing technology.
- *Piezoelectric sensors*: These sensors measure dynamic responses such as vibration and can detect structural abnormalities like excessive vibration due to traffic or wind loads, as well as measuring dynamic characteristics of a structure such as its natural frequencies, modal shapes and damping. For example, Miccinesi et al. [5] investigated the structural response of a reinforced concrete tower using this type of accelerometer to assess natural frequency and wind-induced vibrations.
- *Fibre-optic sensors*: These sensors are capable of long-term, real-time monitoring of strain and temperature, and they are ideal for environments where other sensors might deteriorate, such as in bridges exposed to harsh weather. An excellent example of their use is the monitoring of the Tsing Ma Bridge in Hong Kong, where fibre-optic sensors have been employed to continuously monitor strain in the cables and deck elements [6].
- *Sensors measuring changes in electrical resistance, capacitance, and inductance*: These are often integrated into structural elements like steel reinforcement bars to measure stress or strain and are typically used in conjunction with other sensing technologies to enhance data reliability.
- *Photogrammetry-based sensors*: These are non-contact, image-based sensors that use multiple camera setups to measure displacement and deformation. Photogrammetry was successfully used by Zheng [7] to measure structural deformation, achieving millimetre-level accuracy comparable to traditional total station measurements.

These sensors connect to data acquisition systems that store measurements digitally on computers. The connection between sensors and data loggers can be either wired or wireless. Wireless sensor networks (WSNs) have become increasingly popular due to their ease of installation and ability to cover large areas with minimal disruption to the structure's operation.

When designing a monitoring system, decisions about sensor technology, sampling rate (how often data is recorded), and system design are based on both technical and economic factors (refer to sample questions provided for decision making regarding the design of a monitoring system in Section 2.6). Engineers must balance the trade-offs between sensor accuracy, durability, and cost while ensuring the system's ability to function in various environmental conditions.

National and international standards guide the use of advanced monitoring tools. For example, according to Publication 139 of the Iranian Bridge Loading Code [8] (the official standard for design and evaluation of bridge structures in Iran), for field load testing of road bridges with spans under 40 m, mid-span and support displacement measurements during static loading are sufficient. Measurements must be accurate to one-hundredth of the maximum displacement.

The least expensive method that meets these accuracy requirements without disrupting traffic is surveying with a total station. However, in cases where more detailed dynamic behaviour must be captured, other methods such as laser displacement sensors or photogrammetry-based systems are recommended. For railway bridges, dynamic load tests (with speeds of at least 20 km/h) are required. Tools like linear variable differential transformer (LVDT) sensors, laser displacement sensors, or photogrammetry-based systems are better suited for capturing displacement data dynamically, as total stations are inadequate for this purpose.

To complement displacement measurements, strain gauges and accelerometers are commonly used. Strain gauges measure stress at specific points, allowing the calculation of forces (bending moments, axial forces, or shear forces) across a structural section. Resistance strain gauges are widely used in this field because of their simplicity and reliability. An excellent case study of their application is the monitoring of the Humber Bridge in the UK, where strain gauges were installed to assess the effects of traffic and wind loads on the suspension cables [9].

Accelerometers measure dynamic characteristics such as natural frequencies, damping ratios, and mode shapes. When installed at multiple points, they allow the determination of both natural frequencies and the structural mode shapes. For example, during the monitoring of the Jindo Bridge in South Korea, accelerometers helped identify shifts in the bridge's natural frequencies, which indicated a gradual reduction in the stiffness of the bridge [10].

3.4 STRAIN MEASUREMENT SENSORS

Strain gauges are among the most commonly used sensors for measuring strain in structural elements. Typically, these gauges are resistance-based sensors (such as the standard 120-ohm strain gauge). They function by detecting changes in the electrical resistance of the gauge, which occurs as the structural element to which the gauge is attached deforms under load. These variations in resistance are minuscule and require a Wheatstone Bridge circuit to detect them accurately (Figure 3.2). The Wheatstone Bridge circuit can be configured in several ways, depending on the number of strain gauges mounted on the structure:

- *Full-bridge configuration*: All four resistors in the Wheatstone Bridge are strain gauges installed directly on the structure (Figures 3.2 and 3.3). This configuration provides the highest sensitivity and is commonly used for precision applications where a comprehensive measurement of strain is required as it enables compensation for temperature variation. For example, full-bridge configurations are often employed in fatigue testing of steel bridge members (Figure 3.4).
- *Half-bridge configuration*: Two strain gauges are installed on the structure, while the remaining two resistors are located in the data acquisition system (Figure 3.5). This method is commonly used for measuring bending moments.

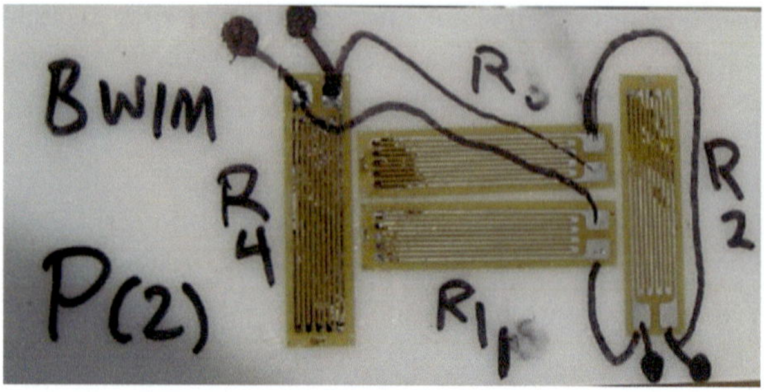

FIGURE 3.2 Full-bridge Wheatstone circuit with four strain gauges installed.

FIGURE 3.3 Full-bridge strain gauge installation on a rail web to measure longitudinal stress.

- *Quarter-bridge configuration*: This setup uses a single strain gauge mounted on the structure, with the remaining three resistors housed in the data acquisition system (Figure 3.6). It's a more economical option, often employed in short measurements where the location of interest is known and high precision is not critical.

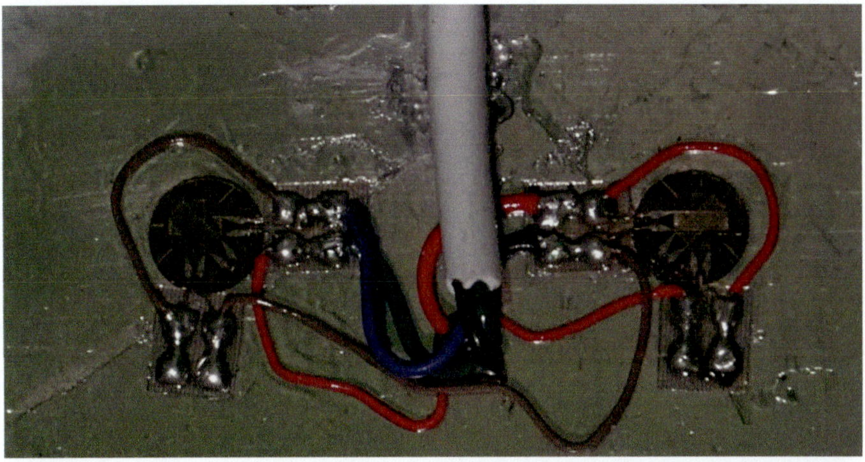

FIGURE 3.4 Full-bridge strain gauge installation on a steel plate girder.

FIGURE 3.5 Rosette strain gauge installed on a rail.

Installing strain gauges demands significant care and expertise. The surface of the structural element must be carefully prepared to ensure strong adhesion, typically involving grinding and sanding to create a smooth, clean surface. Once prepared, the strain gauge is bonded to the structure using a specialized adhesive (Figure 3.7), after which the installation is tested for short circuits using a voltmeter. Lastly, the gauge is soldered into the electrical circuit (Figure 3.8).

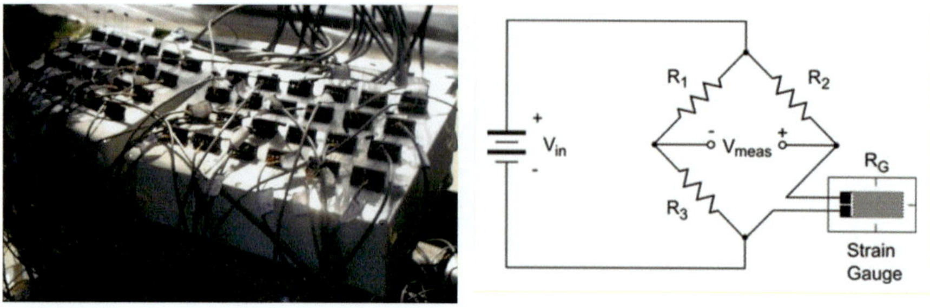

FIGURE 3.6 Quarter-bridge Wheatstone circuit with one strain gauge and connection box to data logger.

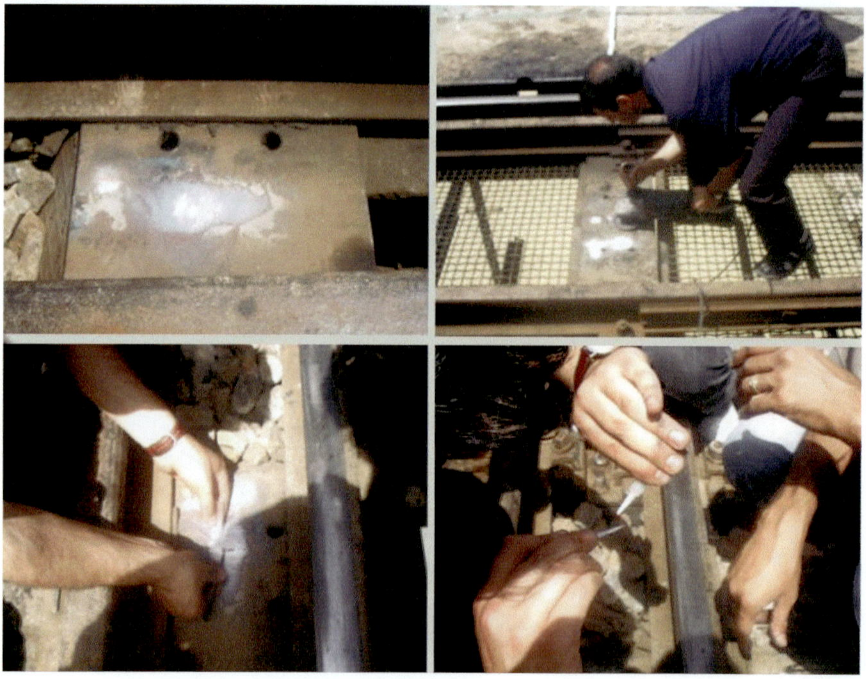

FIGURE 3.7 Grinding, sanding, and bonding of the strain gauge on the upper flange of a beam section.

Strain measurements reflect the average deformation over the length of the gauge. For concrete structures, where micro-cracks in tension zones can influence localized strain measurements, longer gauge lengths (on the order of several centimetres) are generally preferred (Figure 3.9). In contrast, shorter gauges (under 1 cm in length) are typically used for steel structures, as they provide higher resolution for detecting localized strain variations (Figure 3.10).

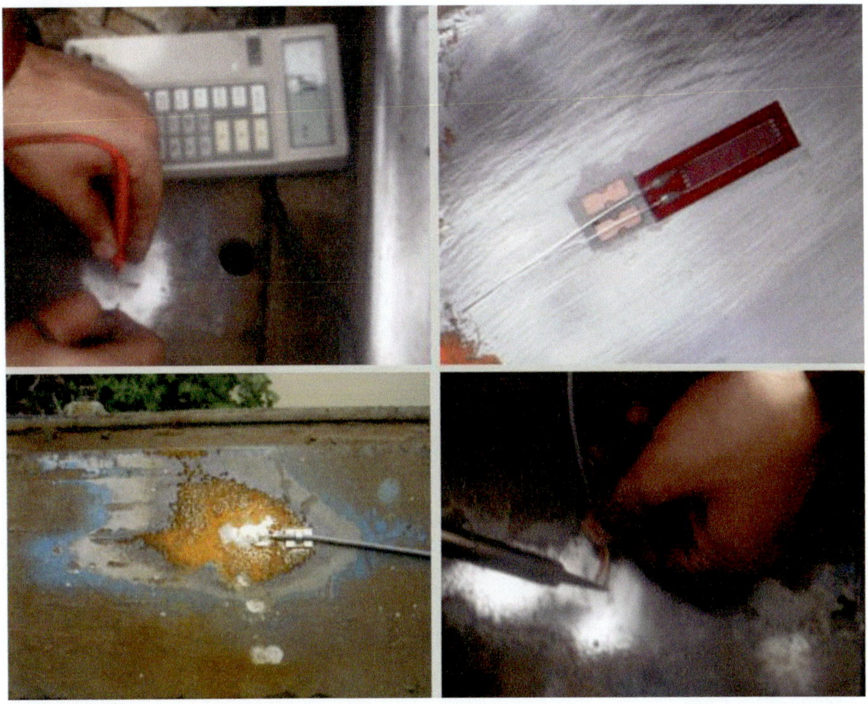

FIGURE 3.8 Checking for short circuits and soldering the strain gauge cable to the electric circuit.

FIGURE 3.9 Strain gauge installed on concrete.

FIGURE 3.10 Strain gauge installed on a rail and steel sleeper in a quarter-bridge configuration.

3.5 MONITORING THE STRAIN IN A RAILWAY BRIDGE

Strain gauges are vital instruments in the field of structural health monitoring, particularly for evaluating the integrity of exposed elements that may experience localized stress concentrations. These stress concentrations often arise in structural components with reduced cross-sectional areas, which can be exacerbated by factors such as corrosion, fatigue, or physical damage. In such scenarios, strain gauges provide critical data on the stresses that develop within these components, allowing engineers to make informed decisions about necessary interventions or reinforcements.

When evaluating the structural capacity of an old concrete slab bridge, strain gauges become an indispensable tool. This old bridge, which originally designed to accommodate axle loads of 20 tons, underwent a change in design parameters necessitating an increase in load-bearing capacity to 22.5 tons. Such upgrades are not uncommon in ageing infrastructure, where evolving operational requirements necessitate enhancements.

To address this increased load requirement, external reinforcements (steel belts) were strategically installed. These reinforcements are critical for distributing the increased loads and mitigating potential failure points. Strain gauges were affixed to these steel belts to measure the effectiveness of the reinforcements during actual loading conditions.

To assess the effectiveness of these installations reinforcements, it is crucial to conduct controlled tests where trains (or other vehicles) cross over the bridge under expected loading conditions. The data from the strain gauges can then be analysed alongside numerical models that simulate the expected performance of the reinforced structure. This comparative analysis allows engineers to validate the design and ensure that the reinforcement behaves as intended under operational conditions.

In the case study mentioned, the installation of strain gauges on the external steel belts provided strain data during loading tests. Figure 3.11 shows a strain gauge mounted on a reinforcing belt within a 4-metre-span concrete slab. This setup allows for localized measurements that provide insights into how the reinforcement behaves under varying loads. Figure 3.12 illustrates the load arrangement, which is critical for understanding the context of the measurements taken. The recorded strain from the gauges, as depicted in Figure 3.13, demonstrates a maximum strain of 60 μstrains, translating to a stress of 12 MPa.

These findings were then compared with the predictions generated by the numerical model developed for the bridge. The analysis confirmed that the strain responses were within acceptable limits, leading to the conclusion that the bridge could safely accommodate the increased load of 22.5 tons. This outcome not only validated the effectiveness of the reinforcements but also instilled confidence in the continued operation of the bridge under revised loading conditions.

FIGURE 3.11 Installing a strain gauge on the reinforcing steel belt of a concrete bridge.

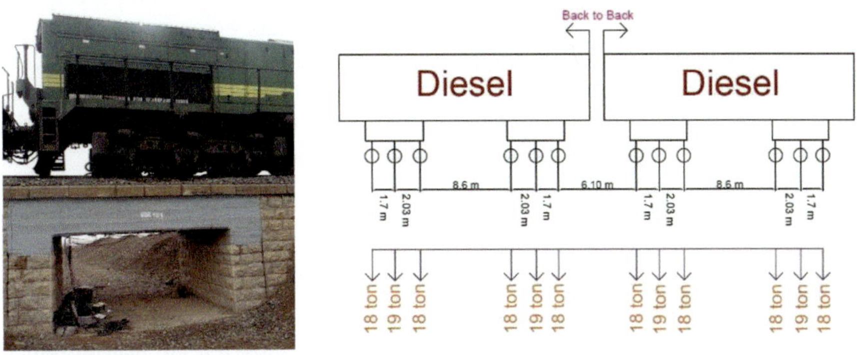

FIGURE 3.12 Train setup during the field testing of the bridge.

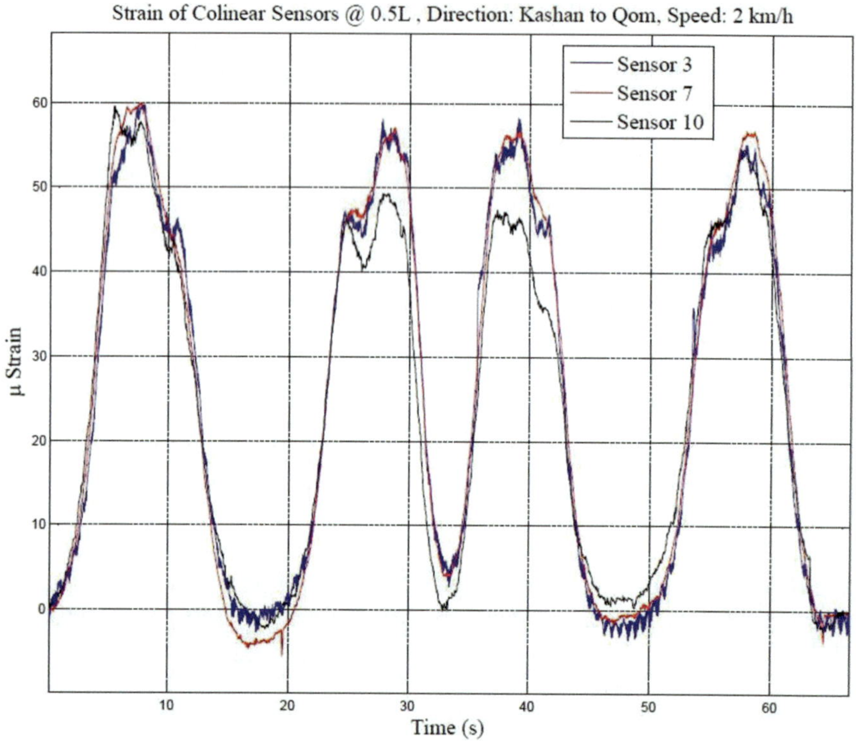

FIGURE 3.13 Strain signature on the reinforcing belts under test train loading.

3.6 RELATIVE DISPLACEMENT MEASUREMENT SENSORS

Relative displacement measurement is crucial when tracking movements between structural elements, such as the widening of cracks, shifts between stone blocks in masonry structures, longitudinal displacements of bridge decks, or vertical displacement in railway sleepers. Traditionally, these measurements were taken using simple mechanical methods, such as sealing a crack with glass and gypsum (Figure 3.14). If the glass cracked, it would signal relative movement at the location. While such methods were widely used in older masonry structures, they only provided qualitative results and could not quantify the magnitude of displacement.

For more accurate and quantitative assessments, strain gauges (Figure 3.15), LVDTs (linear variable differential transformers) (Figure 3.16), or load cells (Figures 3.17 and 3.18) are used. Strain gauges offer the highest level of accuracy among these devices, making them ideal for tracking minute displacements.

In cases where a larger measurement range is required, LVDTs are often employed. These devices measure displacement between two points, converting physical movement into a proportional electrical signal. Load cells, which measure the force exerted on a structural element, can also be used in certain displacement monitoring applications.

FIGURE 3.14 Sealing a crack with glass and gypsum (traditional monitoring).

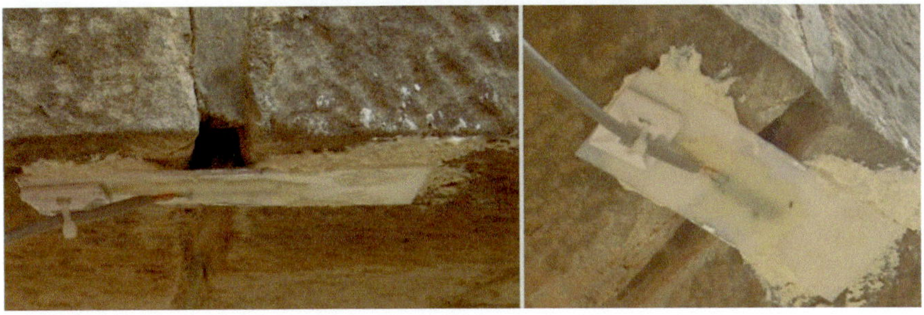

FIGURE 3.15 Measuring relative displacement between two stone blocks by a strain gauge.

3.6.1 Monitoring Relative Displacement between Two Masonry Blocks under Train Load

During the inspection of a masonry arch bridge, a dislodged stone block was observed, suggesting potential structural deterioration. The initial remediation involved grouting the voids created by the dislodgement, which temporarily restored the bridge's integrity. However, upon subsequent inspections, the formation of cracks in the vicinity of the repaired area raised concerns about the effectiveness of the repair and the overall health of the structure. The cracks indicated that underlying structural issues, such as uneven load distribution or material fatigue, might still exist. Consequently, the asset manager sought to gain deeper insights into the structural behaviour of the bridge, specifically how it responded to dynamic loads caused by passing trains.

FIGURE 3.16 Measuring crack widening with an LVDT.

FIGURE 3.17 Measuring relative displacement with a 50 cm gauge by load cell on a concrete beam.

FIGURE 3.18 Measuring longitudinal displacement of a steel beam relative to an abutment using a load cell.

To achieve this, the decision was made to develop a complete three-dimensional (3D) numerical model of the bridge. Numerical modelling, often using finite element methods (FEM), allows engineers to simulate the bridge's response to various loads and boundary conditions, replicating real-world operational scenarios. By incorporating the geometry and material properties of the bridge, the model aimed to predict stress distributions, deformations, and potential failure mechanisms under train loads. Load rating calculations were also performed to determine the maximum allowable train loads, ensuring the bridge's safety for continued operation.

Given the unique characteristics of masonry structures, which are highly dependent on the properties of the individual stone blocks and mortar, several material samples were collected from the bridge for laboratory testing. These tests helped define the mechanical properties of the stone, such as compressive strength, modulus of elasticity, and Poisson's ratio, which are critical inputs for the accuracy of the numerical model.

To further understand the bridge's behaviour, a monitoring campaign was conducted. The bridge was instrumented with various sensors to capture its global and local response to train loads. Strain gauges were installed to measure strain in the masonry blocks, providing insights into stress distribution.

Monitoring this local behaviour was useful in understanding whether the cracks observed during inspection were due to ongoing relative movement between the

FIGURE 3.19 Installing a strain gauge sensor to monitor relative displacement between masonry blocks.

blocks. Figure 3.19 illustrates the installation of this strain gauge between the two masonry blocks. The recorded data helped detect minute shifts or separations, which, if present, could indicate progressive deterioration or insufficient bonding from the previous grout repair. The results of these measurements are presented in Figure 3.20.

Incorporating such localized details, such as relative displacement between individual stone blocks, into numerical models is challenging. Masonry bridges are often composed of irregular stone blocks, each behaving independently to some extent. Traditional FEM struggles to capture this discrete behaviour, as it assumes continuous material behaviour. Therefore, advanced techniques, such as discrete element modelling (DEM), are better suited for modelling masonry structures in cases where individual blocks and their interactions (e.g., friction, separation, rotation) need to be simulated. DEM allows for the explicit representation of each stone block, providing a more accurate simulation of how the structure might behave under real-world conditions, particularly when individual elements of the masonry can shift, crack, or break apart.

3.7 DISPLACEMENT MEASUREMENT SENSORS

The field measurement of displacement is a fundamental technique in SHM, providing direct insights into the deformation of a structure under loads. Displacement data is crucial for assessing the integrity and serviceability of a structure, as well as understanding how it responds to various types of stressors such as traffic loads, thermal expansion, or environmental factors like wind.

Static displacement measurements are typically conducted through surveying techniques. A common method involves using a total station, a device capable

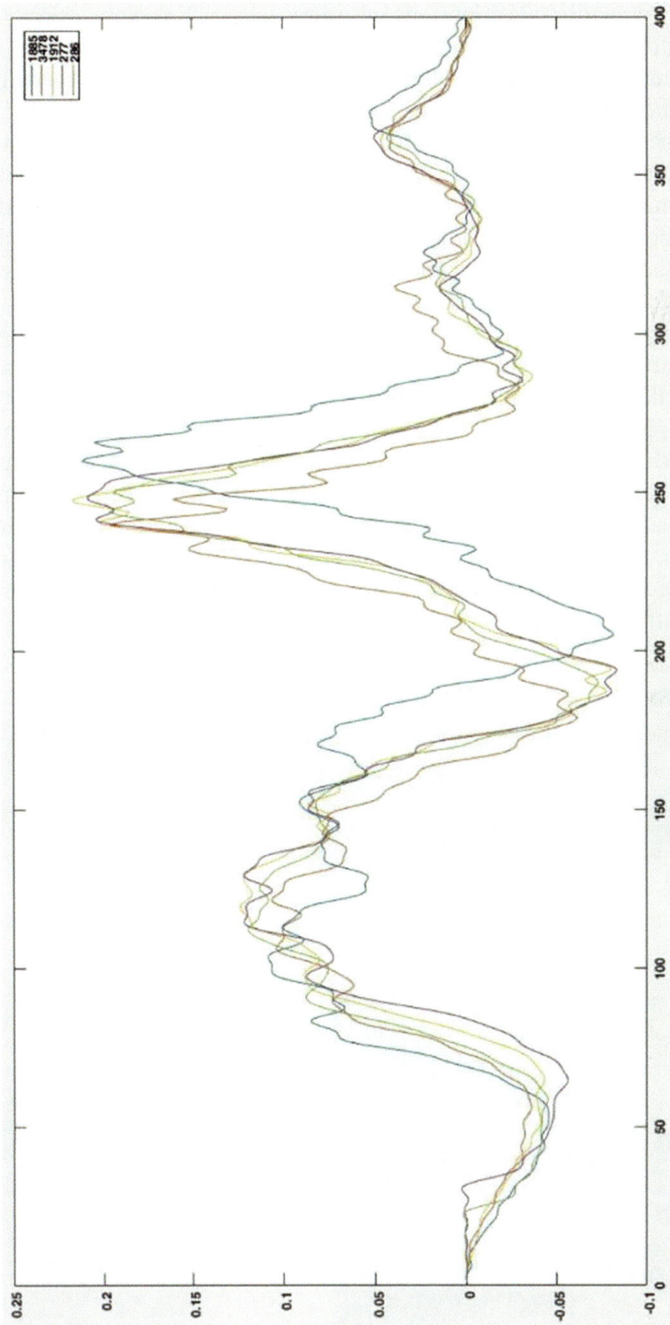

FIGURE 3.20 Relative displacement between two masonry blocks under train load.

of tracking the displacement of specific points on the structure (Figure 3.21). This approach is widely used due to its cost-effectiveness and simplicity of deployment. However, it is primarily suited for static measurements, such as those needed to assess long-term deformations caused by environmental conditions or slow structural settling.

The accuracy of this method largely depends on the distance between the total station and the monitored points. The farther the distance, the higher the likelihood of measurement errors due to line-of-sight obstruction, atmospheric conditions, or mechanical limitations of the total station itself.

For dynamic displacement or time history measurements, linear variable differential transformers (LVDTs) are commonly employed. LVDTs are transducers that convert the displacement of an object into a corresponding electrical signal. They require a fixed reference point to measure the displacement, which makes them particularly suitable for applications such as monitoring the vertical displacement of bridge decks under live traffic loads (Figures 3.22 and 3.23). LVDTs offer high accuracy and real-time data, making them a preferred choice for dynamic monitoring.

For example, LVDTs were extensively used in the SHM of the Humber Bridge in the UK, where they measured the vertical displacement of the deck under traffic and wind loads. Data from the LVDTs allowed engineers to assess the impact of these loads and make decisions regarding structural retrofitting and maintenance [9]. However, the installation of LVDTs on large or tall structures can present logistical challenges, as the sensors need to be stationary relative to the structure, often requiring scaffolding, cranes, or other support systems.

FIGURE 3.21 Static displacement measurement of a bridge using surveying.

FIGURE 3.22 Scaffolding setup for measuring bridge displacement with an LVDT.

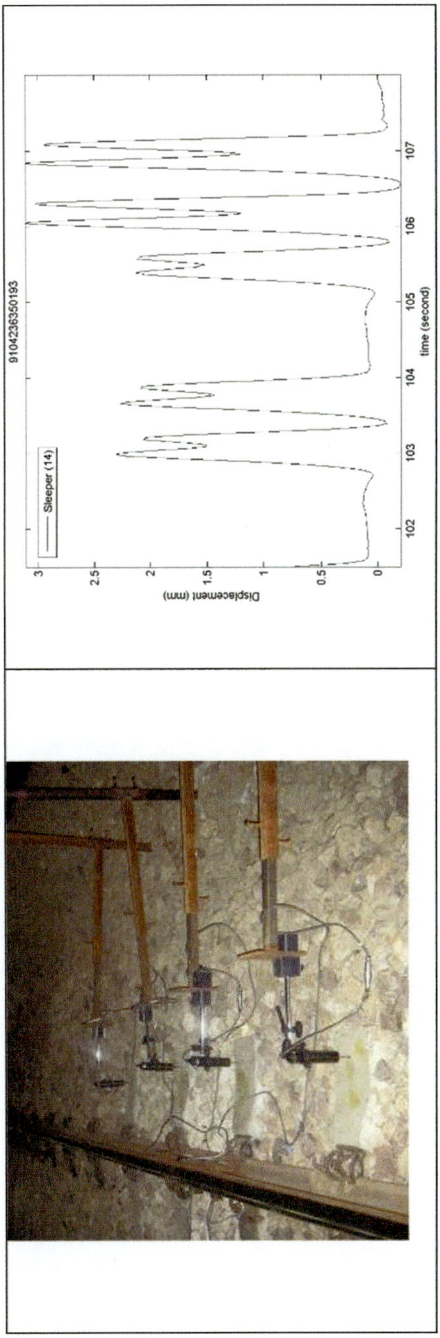

FIGURE 3.23 Measuring vertical displacement of sleepers with LVDTs under train load: (a) sensor setup and (b) common results.

In environments where LVDT deployment is not feasible, such as on tall structures or structures where a stable reference point is difficult to establish, Differential capacitive displacement transducers (DCDTs) offer an alternative. DCDTs measure the stretching of a wire (up to 20 m in length) between the sensor and a fixed ground point (Figures 3.24 and 3.25). This method is particularly effective in historical structures like masonry bridges, where installing sensors at the mid-span or attaching equipment directly to the structure may be impractical or undesirable due to preservation concerns.

However, DCDTs can be sensitive to environmental factors such as wind, which can cause the wire to vibrate, introducing noise and inaccuracies in the displacement

FIGURE 3.24 Measuring displacement of a masonry arch bridge with a DCDT.

FIGURE 3.25 Triangulation-based displacement measurement for a high pier bridge.

FIGURE 3.26 Measuring displacement of the Ghotour bridge using DIC.

data. To mitigate these challenges, it is advisable to conduct measurements under calm weather conditions and ensure that the wire is properly tensioned to reduce oscillations.

In situations where it is impractical or invasive to install physical sensors, non-contact measurement techniques such as photogrammetry and digital image correlation (DIC) are increasingly popular. These methods involve capturing high-resolution images of a structure and using advanced software to track displacement by analysing the movement of points across successive images (Figure 3.26).

Laser displacement sensors are another non-contact option that operates by directing a laser beam at a target surface and measuring the movement of the reflected light. These sensors are particularly useful in monitoring structures where physical access is limited or hazardous, such as in the case of tall towers or offshore platforms.

In addition to direct displacement measurements, displacement can also be derived by double-integrating acceleration measurements taken from accelerometers. This method is particularly useful for monitoring dynamic responses in structures like tall buildings or bridges during seismic events. However, accurate displacement calculation from acceleration data requires sophisticated algorithms, as the process is highly sensitive to noise. Small errors in the acceleration data can lead to significant discrepancies in the displacement calculation over time.

3.8 VIBRATION MEASUREMENT SENSORS

Vibration measurement is essential in the SHM of various structures, including bridges, railway tracks, and buildings. These measurements help detect structural anomalies, assess performance under dynamic loading, and validate structural models. To analyse the dynamic behaviour of structures, two primary types of vibration sensors are commonly used: accelerometers and velocimeters. Each type is suitable for different frequency ranges and applications.

Accelerometers are ideal for measuring high-frequency vibrations, such as those encountered in railway vehicles, bridge decks, or structural components subject to dynamic loading. They convert mechanical motion into an electrical signal that can be analysed to determine the structure's dynamic response. Velocimeters, on the other hand, are more suited for low-frequency vibrations, such as those produced by long-span bridges or ground vibrations caused by passing trains. Velocimeters measure the velocity of vibrations, which is particularly useful in cases where the displacement and low-frequency response of a structure need to be understood.

The selection of an appropriate vibration sensor is critical to ensure accurate and reliable data. When selecting a vibration sensor, one of the most important factors to consider is the sensor capacity, which refers to the maximum vibration level the sensor can measure before becoming saturated. This is particularly important in high-dynamic environments where significant vibration levels may occur.

For instance:

- Dynamic measurements on bridges or sleepers typically require an accelerometer with a capacity of around 2 g. Such sensors provide a balance between sensitivity and the ability to handle the relatively moderate dynamic loads seen in civil structures (Figures 3.27 and 3.28).
- For more extreme conditions, such as measuring the acceleration at the wheelset of railway vehicles, accelerometers with a capacity of 50–100 g are necessary. This ensures the sensor can capture the high accelerations experienced at high speeds and during impacts (Figure 3.29). In these applications, shock and vibration levels may exceed the capacities of standard accelerometers, making robust high-capacity sensors essential.

FIGURE 3.27 Installation of single-axis accelerometers (vertical and lateral) and tri-axial accelerometers.

FIGURE 3.28 Installation of single-axis accelerometers on the centre of a sleeper.

- When monitoring the acceleration at the central piece of railway turnouts, where impacts are more concentrated due to rail geometry and wheel–rail interaction, an accelerometer with a capacity of 100 g or more is recommended (Figure 3.30).

In contrast, for ground vibrations caused by passing subway trains, velocimeters are generally preferred. Subway systems typically generate low-frequency vibrations due to their large masses and low operational speeds. These vibrations can propagate through the ground and affect nearby structures, making velocimeters the ideal choice for such applications (Figure 3.31).

Accelerometers can be classified into two main types based on their measurement capabilities: single-axis and tri-axial. Single-axis accelerometers measure vibrations in a single direction (e.g. vertical or lateral), while tri-axial accelerometers measure vibrations along three orthogonal axes (*X, Y, Z*), providing a more comprehensive understanding of a structure's dynamic behaviour. The proper orientation of these

FIGURE 3.29 Installation of a tri-axial accelerometer at the wheelset of a rail wagon.

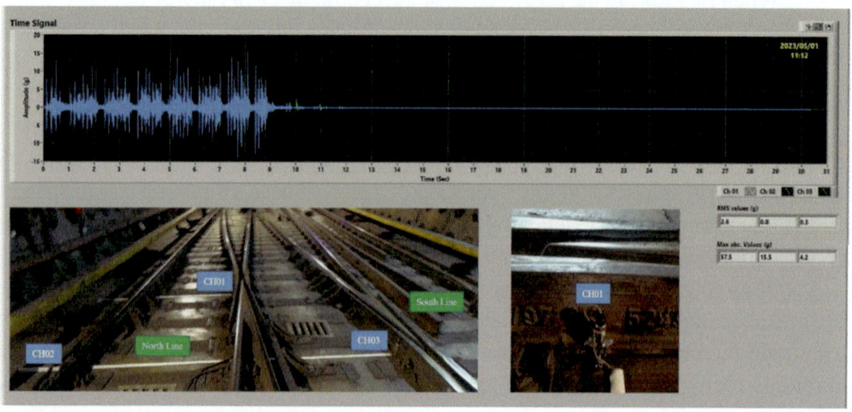

FIGURE 3.30 Installation of an accelerometer to measure vibrations from a train wheel impacting the central section of a turnout.

FIGURE 3.31 Installation of an accelerometer and velocimeter for measuring ground vibrations from passing subway trains.

sensors during installation is crucial to obtain meaningful data, particularly when measuring complex structural responses under dynamic loading conditions.

Calibration of accelerometers is another critical aspect to ensure measurement accuracy. The calibration process involves subjecting the sensor to known vibrational inputs at specific frequencies and comparing the output to a reference sensor. This process helps verify that the accelerometer's output is accurate across its operational frequency range. Regular calibration is recommended, especially in high-stakes environments like railway infrastructure, where safety is paramount. Calibration standards for accelerometers are well-documented in international standards like ISO 16063-21 [11], which provides guidelines on the methods for calibrating vibration and shock transducers.

3.8.1 Monitoring the Condition of a Central Piece of a Turnout in Tehran Metro

Rail turnouts are critical components in railway systems, serving as the points where trains diverge from one track to another. Their design inherently involves discontinuities—where the geometry of the track changes, particularly at rail joints and switches. These discontinuities create a unique challenge: as a train wheel passes over a turnout, it generates a significant impact due to discontinuity. This impact is not merely a momentary event; it can lead to ongoing vibrations that affect the integrity of the track and the safety of train operations.

Measuring the impact generated by wheels passing over turnouts provides two vital insights into the condition of the rail infrastructure:

1. *Track degradation monitoring*: Over time, the repeated impacts can lead to wear and tear on the track, causing degradation at the turnout location. By continuously monitoring the impact forces, asset managers can identify trends indicating deterioration. Early detection of such degradation allows for timely maintenance interventions, potentially preventing costly repairs or accidents.
2. *Wheel roundness deviation assessment*: Comparing the impacts generated by each wheel across an entire train helps in assessing the roundness of the wheels. Deviations from the ideal circular shape can result in uneven forces being applied to the track. Monitoring these deviations not only informs about the state of the wheels but also highlights the necessity for maintenance of the rolling stock. Ensuring that wheels are within acceptable tolerances contributes to the overall safety and reliability of train operations.

Given the frequent defects observed in one of the turnouts in the Tehran Metro, the asset manager initiated a structural health monitoring campaign to investigate the underlying causes of these issues. This proactive approach involved deploying a monitoring system specifically designed for the critical turnout.

As illustrated in Figure 3.30, the monitoring system consists of three strategically placed accelerometers:

- Accelerometer 1: Measures the vertical acceleration at the nose of the central piece of the southern track, capturing the immediate impact forces as a train wheel passes over this critical point.
- Accelerometer 2: Positioned at the rail joint adjacent to the central piece of the northern track, this sensor provides insights into the forces exerted at the interface between two rail segments, a known point of vulnerability.
- Accelerometer 3: Located at the nose of the central piece of the northern track, this sensor complements the data collected by the first two, allowing for a comprehensive understanding of the dynamics at play.

Given the anticipated high-impact loads at this location, all accelerometers were selected with a capacity of 100 g to ensure accurate readings without saturation during peak load conditions.

Figure 3.30 also presents the time history of the vertical accelerations recorded by these three sensors for a train passing on the left track at a speed of 49 km/h. The data generated allows for detailed analysis of the impact forces over time, enabling engineers to:

- Evaluate the performance of the turnout under operational conditions.

- Identify peaks in acceleration that may correlate with specific defects or irregularities in the track or rolling stock.
- Conduct comparative analyses between different train speeds and conditions, further refining the understanding of how various factors influence the structural health of the turnout.

Through this monitoring initiative, asset managers gain invaluable data that aids in maintaining the integrity of the railway infrastructure, ensuring not only the safety of passengers but also the efficiency of the entire metro system.

3.9 DATA ACQUISITION SYSTEMS

To power, amplify, record, and store the data from vibration sensors, data loggers, or data acquisition systems are used. These systems vary in terms of dynamic range, sampling rate, number of channels, and types of sensors they can measure (Figures 3.32 and 3.33). For example, a measurement channel could handle:

- Voltage: for displacement, force, and acceleration measurements.
- Milli-voltage: used with strain gauges requiring a voltage amplifier.

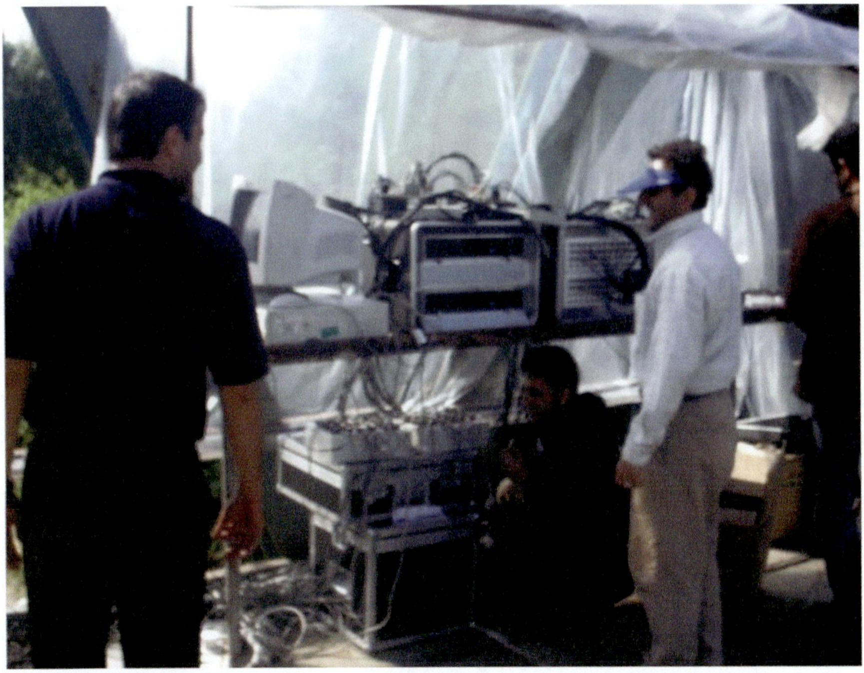

FIGURE 3.32 96-Channel data logger for measuring strain, displacement, and acceleration at the Neka Bridge in 2004.

FIGURE 3.33 16-Channel permanent data logger for strain measurement at the Veresk Bridge in 2016.

Most data loggers feature an analogue-to-digital converter (e.g., 16-bit or 24-bit) which converts the analogue signal into digital units for storage. The sampling rate of the data logger refers to how many samples are collected per second (e.g., 20 Hz, 200 Hz, 2000 Hz). Low sampling rates are used for static data, while higher rates are necessary for dynamic measurements.

3.10 DATA PROCESSING

Once a structure is instrumented with sensors, the raw data generated is usually riddled with noise, which can obscure or distort the true signal. This noise can originate from various sources, including environmental factors (such as temperature fluctuations, wind, or vibrations from nearby machinery), imperfections in the sensors or wiring, and electromagnetic interference. Effective data processing is therefore critical in SHM to ensure that the data accurately represents the structural behaviour without external distortion.

3.10.1 Zero Offset and Trend Removal

Zero offset and trend issues are common in SHM data. Zero offset refers to a situation where a sensor signal reads a value other than zero in a state where zero is expected (e.g., no load or strain in a system). In structural monitoring, such offsets can arise from calibration issues, environmental drift, or other sensor-related anomalies.

To address zero offset, a typical correction involves shifting the data so that the initial measurement (or an average of the initial set of measurements) represents zero. Mathematically, if $y(0)$ is the first value in a set of measurements $y(t)$, each subsequent value $y(t)$ can be adjusted by subtracting $y(0)$ to eliminate the offset.

Another issue in SHM data is the presence of trends that can appear as a linear or nonlinear progression in the signal. Trends may result from factors such as temperature changes over time or slow drifts in the sensor baseline, which could be mistaken for structural deformation if not addressed. To correct for linear trends, a line fitting the initial and final measurements can be calculated. This process, often referred to as detrending, involves subtracting the values of this fitted line from the original signal at each time point. Polynomial fitting techniques can also be applied to account for nonlinear trends, allowing more complex drift patterns to be removed [12]. Figure 3.34 shows an unfiltered and a filtered data with both trend removal and zero offsetting.

3.10.2 Noise Reduction

A more advanced approach to remove high frequency noises, is to use low-pass filtering, which eliminates high-frequency components associated with measurement. Low-pass filters are effective in retaining the lower-frequency content of the signal that often corresponds to structural responses while attenuating higher frequencies linked to environmental noises [12]. Filtering can be done by wavelet

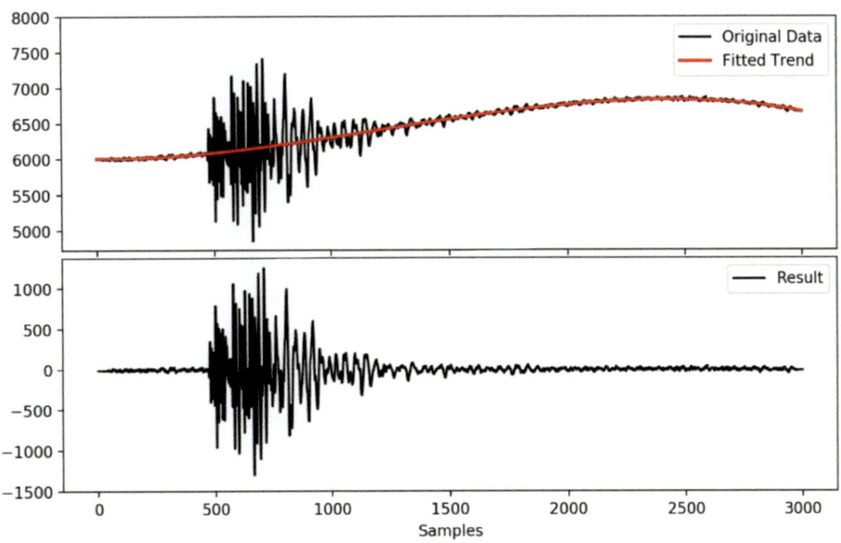

FIGURE 3.34 Zero offset and trend removing in the measured signal.

transform techniques to separate noise and drift from actual structural data, providing an adaptive and localized signal correction method [13].

For instance, in strain measurements, high-frequency noise might appear on the signal from a sensor placed at the mid-span of a transverse beam's lower flange (Figure 3.35). The presence of this high-frequency component results in a maximum unfiltered value of 234.8 μstrain, which could be mistaken for actual structural deformation if not properly addressed.

To mitigate this issue, the signal must be filtered to remove unwanted high-frequency noise while preserving the relevant structural information. Filtering involves the application of a mathematical function or filter that selectively removes certain frequencies from the data. A filter is characterized by its coefficients, typically denoted as *A* and *B*. The A coefficients relate to the current and past output values of the filter, while the *B* coefficients correspond to the current and past input values of the signal. By applying these coefficients, the filter adjusts the measured signal to retain only the desired frequency components [14].

The application of filters is especially important in SHM to clean the signal, as demonstrated in Figures 3.36 and 3.37. In these figures, the presence of noise in the unfiltered signal is clearly visible, but after the application of a noise-removal filter, the signal becomes significantly cleaner, providing more accurate measurements of structural behaviour.

3.10.3 Butterworth Filter

The Butterworth filter, first introduced by British engineer Stephen Butterworth in 1930, is a type of signal processing filter designed to have a maximally flat frequency response in the passband, meaning it does not exhibit any ripples or variations in the frequencies it allows through. This characteristic makes it especially useful when a smooth, undistorted signal is required, which is why it is widely used in applications such as SHM, where maintaining the integrity of the data is critical.

The primary feature of the Butterworth filter is its frequency response, which is designed to be as flat as possible in the passband (the range of frequencies the filter allows to pass through). As the frequency increases beyond the cutoff, the filter attenuates the signal, with the attenuation rate dependent on the filter's order. The higher the order of the filter, the steeper the roll-off beyond the cutoff frequency, allowing for a cleaner separation of frequencies. The filter is typically characterized by two key components: the cutoff frequency and the filter order.

- *Cutoff frequency*: This is the frequency at which the filter begins to attenuate the signal. For a low-pass filter, this means frequencies higher than the cutoff are gradually reduced, while for a high-pass filter, frequencies lower than the cutoff are attenuated.
- *Filter order*: The order of the filter determines how sharply the transition is from the passband to the stopband. A first-order Butterworth filter will have a relatively shallow transition, while higher-order filters (such as second or third order) provide a steeper roll-off.

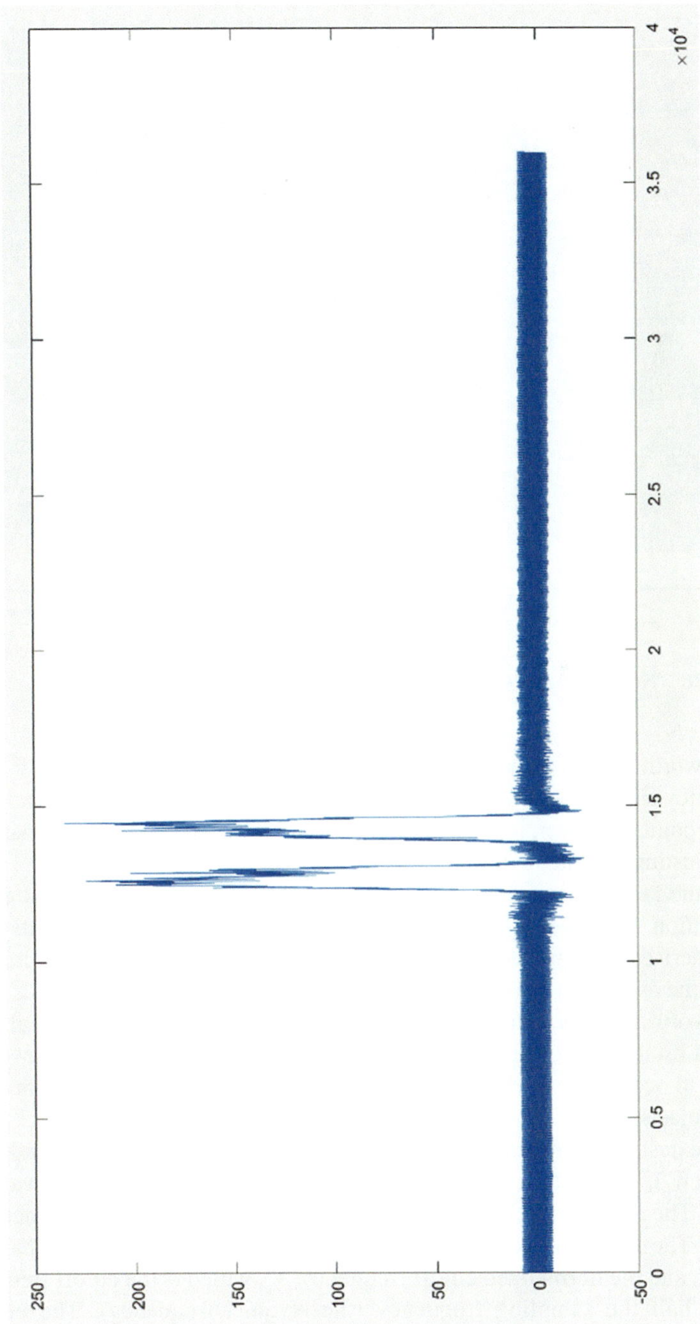

FIGURE 3.35 Presence of high-frequency information (noise) in signal ch45 (strain at mid-span of lower flange of transverse beam).

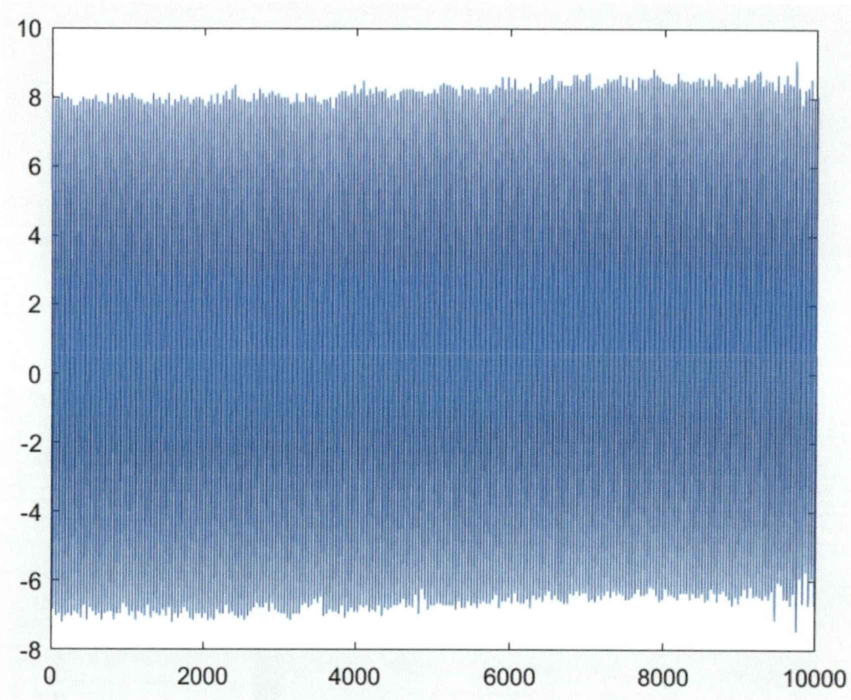

FIGURE 3.36 Noise in signal CH45.

The Butterworth filter is a low-pass filter by default in MATLAB, meaning it allows low-frequency signals to pass and attenuates higher-frequency signals beyond a specific cutoff point. However, it can be modified to function as a high-pass or bandpass filter by adjusting the filter's design.

One of the key advantages of the Butterworth filter is its simplicity and ease of implementation. The filter's transfer function can be defined in terms of its poles, which are derived from the desired cutoff frequency and filter order. The filter does not exhibit the oscillatory behaviour especially near the cutoff frequency. This makes the Butterworth filter ideal for applications where the preservation of signal integrity without oscillations or distortions is paramount, such as in SHM systems where the goal is to accurately track structural changes without introducing errors due to filtering artefacts [15].

Once the design parameters are established, such as the filter order N and cutoff frequency (W_n), the Butterworth filter can be implemented easily in software like MATLAB. The filter coefficients are then computed using the 'butter' function in MATLAB. The function takes the filter order N, which determines the sharpness of the filter, and the normalized cutoff frequency W_n, which is the cutoff frequency relative to half the sampling frequency (the Nyquist frequency). The Nyquist

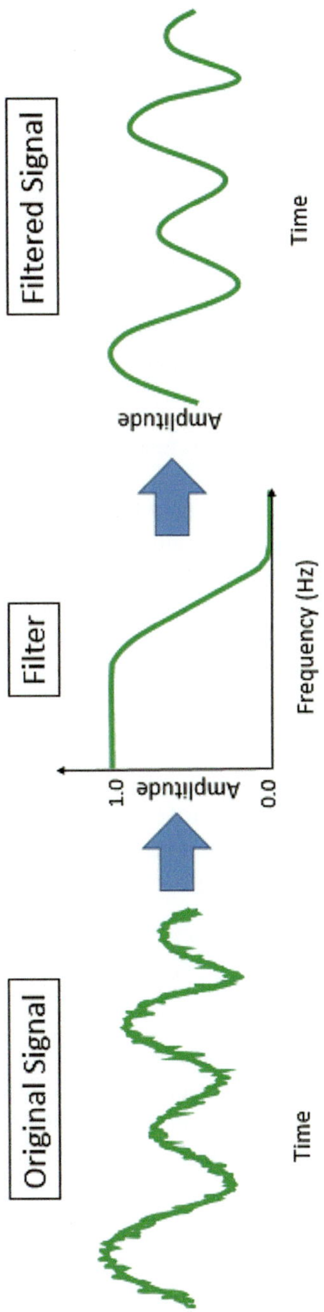

FIGURE 3.37 Noise removal by filtering the signal.

frequency is defined as half of the sampling frequency, which is the rate at which data is collected. For instance, in dynamic bridge measurements, if data is sampled every 0.01 s (sampling frequency=100 Hz), the Nyquist frequency would be 50 Hz.

The general formula for a digital filter that computes the output y_n based on the current and past input values x_n and the previous outputs y_n is given by:

$$a_0 y_n = \sum_{k=0}^{N_b} b_k \, x_{n-k} - \sum_{l=1}^{N_a} a_l \, y_{n-l} \tag{3.1}$$

where

b_k are the filter coefficients for the input signal,

a_l are the filter coefficients for the output signal,

N_b and N_a are the respective order parameters for the numerator and denominator of the filter.

In MATLAB, several type of filter can be created as follows:

- *Low-pass filter*: This filter allows only frequencies below a certain cutoff frequency to pass through. It is commonly used when high-frequency noise needs to be removed, as is often the case in SHM data where structural responses are expected to occur at lower frequencies. In MATLAB, a low-pass Butterworth filter can be created using the following command:

$$[B, A] = butter(N, Wn, 'low') \tag{3.2}$$

- *High-pass filter*: This filter allows only frequencies above a specified cutoff to pass through. It is useful when low-frequency drift or trends need to be removed, such as those due to environmental factors like temperature changes. To create a high-pass Butterworth filter in MATLAB, the following command can be used:

$$[B, A] = butter(N, Wn, 'high') \tag{3.3}$$

- *Bandpass filter*: This filter allows only frequencies within a certain range (between two cutoff frequencies) to pass through. It is used when the relevant signal lies within a specific frequency band, and other frequencies, both higher and lower, need to be suppressed. The MATLAB command for a bandpass filter is:

$$[B, A] = butter(N, [\mathrm{Wn1},\ \mathrm{Wn2}]) \tag{3.4}$$

Once the filter coefficients A and B are determined using the butter function, the next step is to apply the filter to the measured signal. In MATLAB, this can be done using the filter function, which applies the filter coefficients A and B to the input signal x, producing the filtered output signal. The basic syntax for applying the filter is as follows:

$$xf1 = filter(B, A, x) \tag{3.5}$$

This function works by processing the signal in a forward direction, meaning it applies the filter sequentially from the first to the last sample.

Alternatively, MATLAB provides the filtfilt function, which can be used to apply the filter in both forward and reverse directions. The advantage of using filtfilt is that it eliminates phase distortion, which is a common issue when filtering signals in only one direction. Phase distortion occurs because filtering changes the timing of the signal components, especially near the cutoff frequency. By applying the filter in both directions, filtfilt ensures that the output is phase-corrected, meaning that the filtered signal retains its original timing and alignment with the input data. The syntax for using filtfilt is:

$$xf2 = filtfilt(B, A, x) \tag{3.6}$$

Both methods are useful depending on the requirements of the specific SHM application. However, when preserving the timing and phase relationship of the structural signals is critical, the 'filtfilt' function is preferred.

Figure 3.38 illustrates the frequency response (both magnitude and phase) of a fourth-order Butterworth bandpass filter with a normalized cutoff frequency range of [0.04, 0.12], showing how the filter attenuates frequencies outside this range and preserves the desired frequencies within the passband.

3.10.4 Moving Average Filter

In addition to the Butterworth filter, another commonly used filter for smoothing signals is the moving average filter. This filter is particularly useful for reducing noise by averaging a set number of data points around each sample, effectively smoothing out fluctuations that are not part of the underlying trend. The moving average filter is simple to implement and can be effective for situations where high-frequency noise is present but the overall trend of the data is of primary interest.

The 'filter' function in MATLAB can be used to apply this moving average filter to the input signal x, which results in the filtered output signal xf3:

$$xf3 = filter(B, A, x) \tag{3.7}$$

For example, the unfiltered signal in Figure 3.39 has a maximum value of 234.8 μstrain, which is significantly reduced to 195.1 μstrain after applying the 10-point

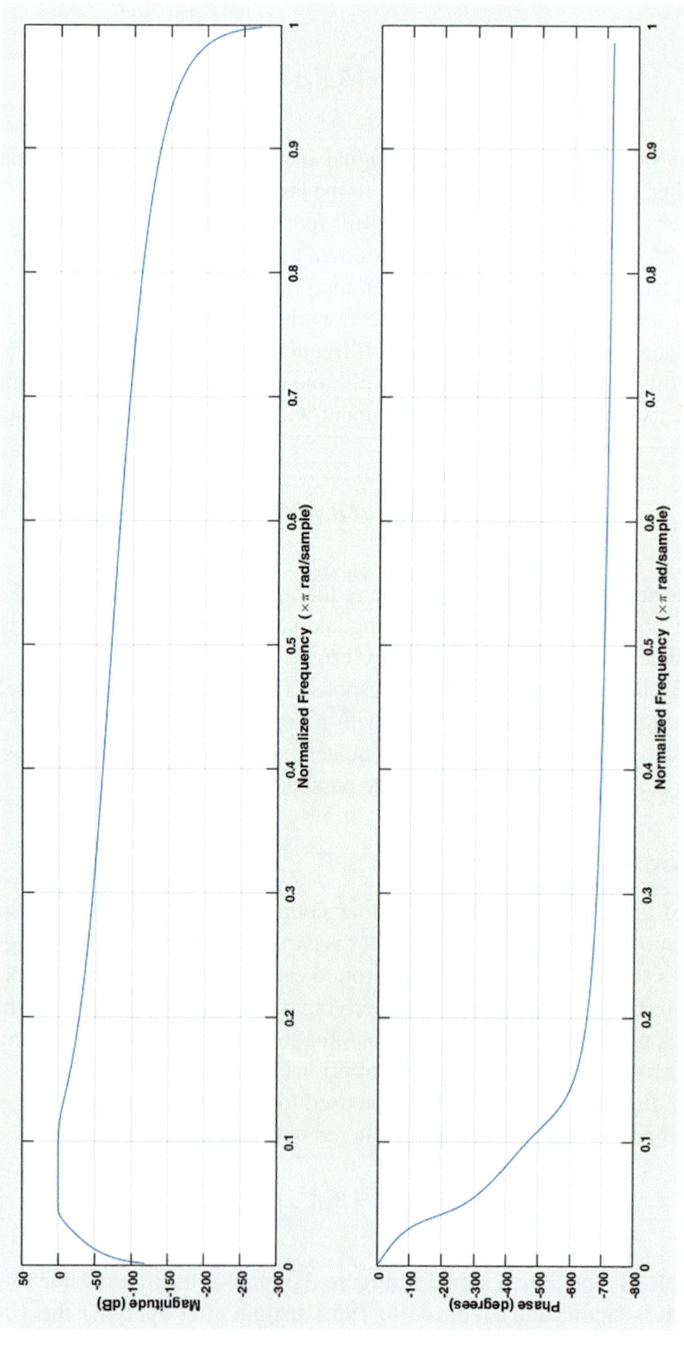

FIGURE 3.38 Frequency response (magnitude and phase) of a fourth-order Butterworth bandpass filter with normalized cutoff frequency [0.04, 0.12].

moving average filter. This illustrates how the filter smooths out high-frequency noise, allowing for a clearer representation of the structural response.

3.10.5 Resampling of Signal

In structural health monitoring, the resampling is useful when there is a need to reduce the sampling rate of a signal to manage data size or to match the sampling requirements of a specific analysis. Resampling involves changing the frequency at which data points are recorded, which can help optimize data storage and processing, especially in long-term monitoring applications where high-frequency data may not be necessary at all times.

The resample function in MATLAB allows for flexible resampling of a signal by adjusting the sampling rate to a new frequency that is a fraction of the original. The resampling factor is defined by the ratio p/q, where p and q are integers that scale the original sampling rate to the desired new rate. The resulting resampled signal x_r is computed as follows:

$$x_r = resample(x, p, q) \tag{3.8}$$

For example, consider a signal with an original sampling rate of 2000 Hz. To downsample this signal to a new rate of 40 Hz, we can use the following command in MATLAB:

$$x_r = resample(x, 40, 2000) \tag{3.9}$$

In this case, setting p=40 and q=2000 reduces the sampling rate by a factor of 50, achieving the desired frequency of 40 Hz without introducing a time delay or phase distortion. This is particularly important in SHM applications, where preserving the timing of the signal is often critical.

Resampling a signal also involves applying an internal low-pass filter, which helps prevent aliasing by attenuating any frequency components higher than half of the new sampling rate (the Nyquist frequency for the resampled signal). This ensures that high-frequency noise or unwanted signal components do not interfere with the lower sampling rate data.

Figure 3.40 demonstrates the difference between the original unfiltered signal at a 2000 Hz sampling rate and the same signal downsampled to 40 Hz, illustrating how resampling effectively reduces the data volume while retaining the essential structural response information.

3.11 LOAD TESTING OF NEKA BRIDGE

To assess the structural health of the Neka Bridge, a monitoring campaign was designed to capture and evaluate the bridge's behaviour under operational loads.

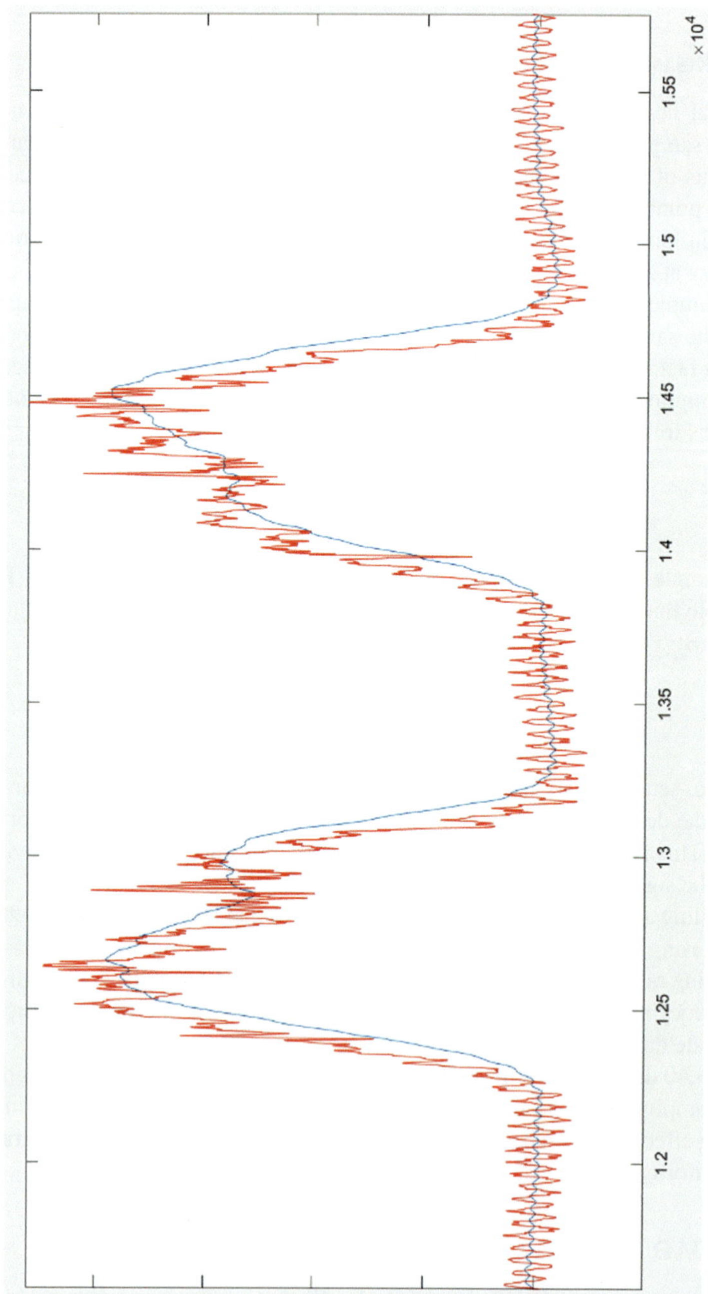

FIGURE 3.39 Unfiltered and filtered signal CH45 (strain at mid-span of lower flange of transverse beam) with moving average filter.

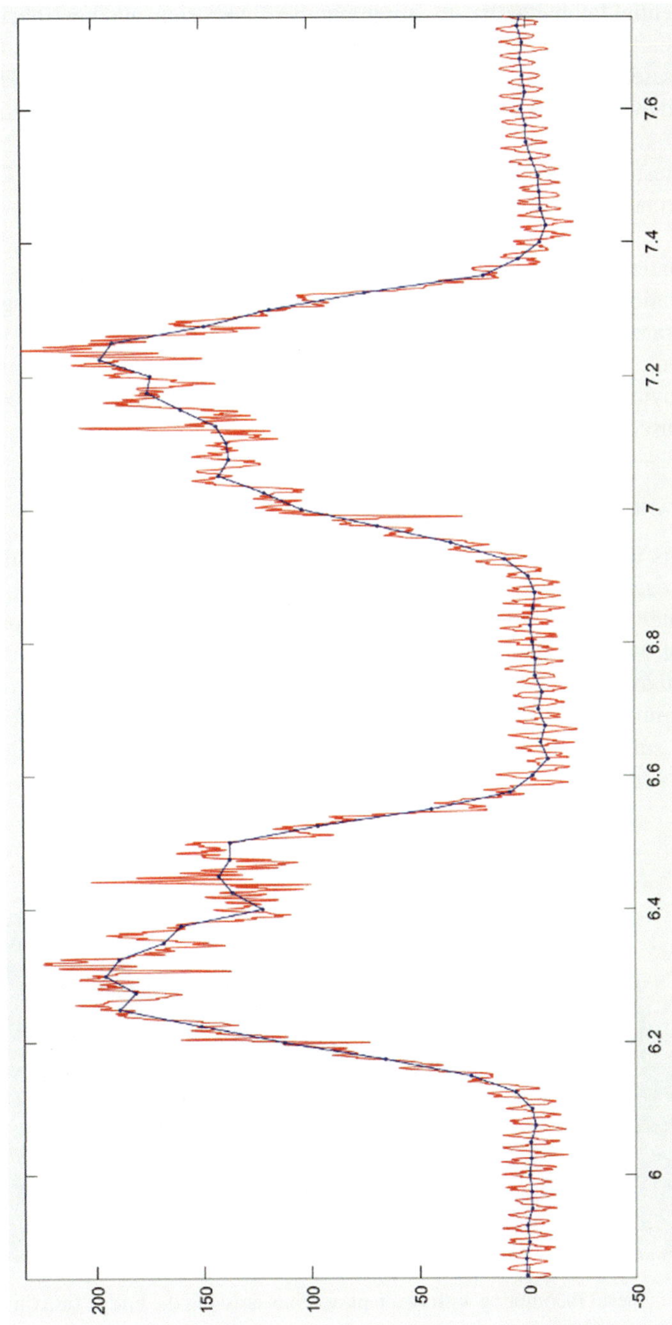

FIGURE 3.40 Unfiltered and sampled signal CH45 (strain at mid-span of lower flange of transverse beam) with re-sampling.

The initial visual inspection, as detailed in the previous chapter, revealed that the superstructure of the bridge appeared to be in poor condition. Signs of deterioration included potential fatigue in the structural members, corrosion, and possible deformation in critical components. These observations raised concerns about the bridge's capacity to safely carry operational loads over its expected service life. Hence, a more detailed evaluation was warranted, which involved both static and dynamic load testing.

The load testing involved both static and dynamic loading of the bridge. Two specific train formations were used to apply these loads:

- Formation 1: A single locomotive.
- Formation 2: A single locomotive accompanied by two loaded wagons (as illustrated in Figure 3.41).

Table 3.1 provides details on the axle configurations and weights of the locomotive and wagons used in these tests.

3.11.1 Static Load Testing

Measurements were taken as the locomotive stopped at specific locations along the bridge, with each stop representing a distinct loading scenario. Table 3.2 shows the eight designated stopping points, with the front axle of the diesel locomotive positioned at intervals of 6 m from the left support. The maximum applied load reached 110.97 tons, when all axles of the locomotive were positioned on the bridge span.

The maximum displacement measured, given the distance from the surveying camera, was sufficient to allow for bridge displacement measurement under static

FIGURE 3.41 Diesel locomotive with two tank wagons crossing the bridge (maximum load applied to the truss: 214.6 tons).

TABLE 3.1
Axle characteristics for locomotive and wagons

Vehicle	Number of Axles	Full Load Weight (tons)	Distance Between Bogie Centres (m)	Distance Between Bogie Axles (m)	Coupler-to-Coupler Distance (m)
GT26 Loco	6	110.97	12.5	2.02–1.69	21.1
Tank Wagon	4	76	9.86	1.8	15.06

TABLE 3.2
Diesel locomotive stopping points during static load testing of the truss bridge

Static Load Stop Stages	1	2	3	4	5	6	7	8
Distance of Front Axle of Diesel Locomotive from Left Support (m)	6	12	18	24	30	36	42	48

loading using the surveying camera. Surveying markers were placed at positions 7.1, 13.4, 20, and 26.2 m from the left corner of the bridge.

The maximum recorded displacement for all eight loading scenarios is presented in Figure 3.42. As shown, the highest displacement occurred at the midpoint of the bridge span, where a maximum deflection of 34 mm was observed.

Static load testing, combined with surveying, is ideal for scenarios where quick and cost-effective measurements of displacement are needed to understand overall bridge behaviour under operational loads. By analysing both the applied loads and the resulting displacements, it is possible to assess the forces within truss members and further refine load rating factors (to be discussed in detail in the next chapter).

3.11.2 Dynamic Load Testing

While static load testing offers valuable insight into a bridge's structural response under operational loads, it does not capture dynamic effects, since the load is applied in a stationary manner. These dynamic effects—such as vibrations, impact forces, and oscillations—are an essential part of how a bridge behaves under real-world conditions with moving loads. Dynamic load testing addresses this limitation by introducing moving loads, allowing for a more comprehensive evaluation of the bridge's behaviour.

However, traditional surveying tools (such as a total station) are not suitable for capturing the rapid, complex responses involved in dynamic load testing. Instead, advanced sensor technology is required to accurately monitor these quick changes

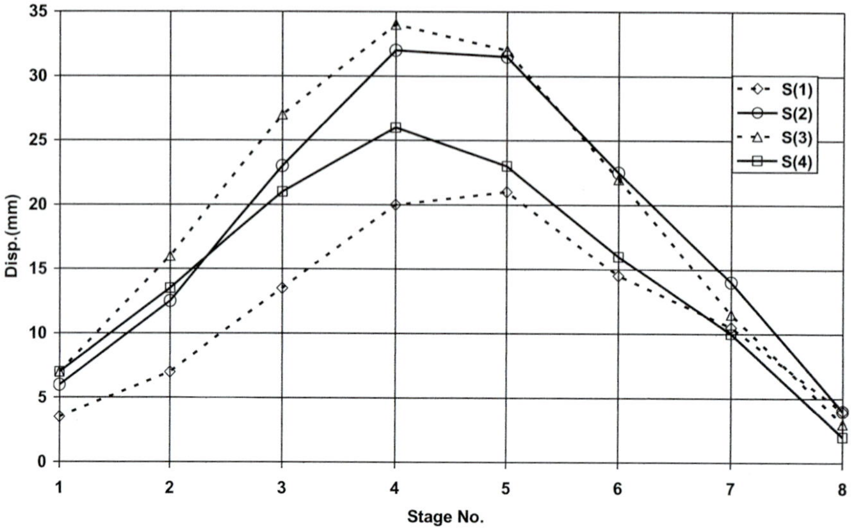

FIGURE 3.42 Vertical displacement of four surveying markers installed on the southern truss during various stages of static loading with a single diesel locomotive, forward movement.

in displacement, acceleration, or strain. Using these sensors, explained in more detail in this chapter, dynamic load testing provides highly reliable measurements of the bridge's response to operational loads, reflecting both the steady and transient behaviours of the structure under real conditions.

3.11.2.1 Vertical Displacement at Mid-span

Figure 3.43 illustrates the maximum vertical displacement under train formation 1 (similar to that used in static load testing) at two different speeds. The maximum vertical displacement is measured at 34 mm, which aligns with the results from static load testing. Furthermore, the results show minimal variation across speeds, suggesting that the dynamic load factor is relatively low.

3.11.2.2 Displacement at Bearings

LVDT results from both static and dynamic tests can be compared to provide insights into the bridge's behaviour under varying loading conditions. Each corner of the truss is supported by a neoprene pad measuring 500×400×100 mm. Visual inspection shows that this neoprene pad consists of six layers, with a combined elastomer thickness of 71 mm. Consequently, each elastomer layer is approximately 11.8 mm thick, yielding a shape factor of 9.4.

During the dynamic test with a single diesel locomotive, the maximum vertical displacement at the right support of the southern truss reached 1 mm, while the lateral displacement was 2 mm (Figure 3.44). In the static test with the same locomotive, the maximum longitudinal displacement at the right support of the southern truss reached 3.3 mm, with a maximum lateral displacement of 2 mm.

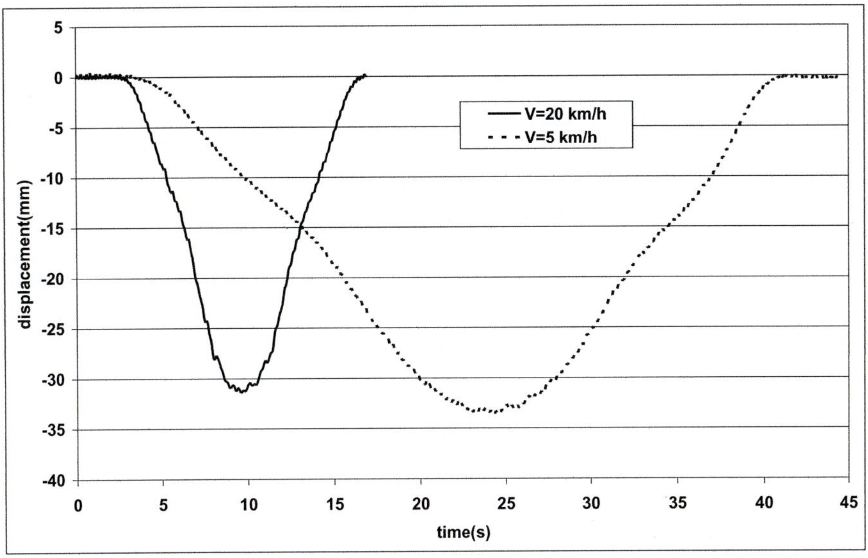

FIGURE 3.43 Vertical displacement at the midpoint of the southern truss span during the passage of train formation 1 at speeds of 5 and 20 km/h.

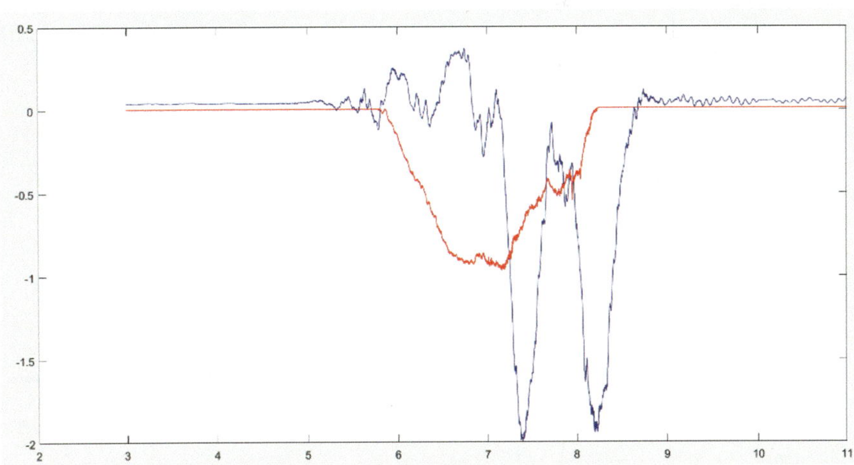

FIGURE 3.44 Lateral and vertical displacement of the right bearing of the southern truss during the passage of train formation 1 (the vertical stiffness of the neoprene is estimated as Kv=27.75 t/mm).

3.11.2.3 Strain Measurements

Several strain gauges were installed on various bridge elements to measure strain values, which will be used to refine the load rating factor (this will be discussed in detail in the next chapter). These gauges are placed on elements expected to exhibit critical strain levels, where load rating assessments and evaluations of the bridge's condition under operational loads are essential.

Figures 3.45 and 3.46 show the strain measured on the longitudinal beams for train formation 1 at speeds of 5 km/h and 20 km/h. The maximum strain in the upper flange of the mid-span longitudinal beam at the midpoint cross-section reached 220 μstrain (equivalent to 44 MPa). This strain range is particularly useful for determining the remaining fatigue life of the steel members, a topic that will be further explored in the following chapter.

Since dynamic load tests were conducted at different speeds, the dynamic load factor can be calculated by dividing the maximum strain observed in each dynamic test by the strain observed in the static test. A histogram of these results can then be used to determine the experimental dynamic load factor, which is typically selected to cover 95% of the observed values.

For the longitudinal beams, the histogram is shown in Figure 3.47. The maximum experimental dynamic amplification factor is 1.34, which is lower than the code-specified value of 1.69 (as shown in the figure's calculations).

Strain measurements for cross-beams, at the top and bottom fibres at the mid-span of the seventh transverse beam, in the upper chord at the sixth panel point of the southern and northern trusses, and strain in the lower chord at the twelfth panel point of the southern and northern trusses are presented in Figures 3.48 through Figure 3.51.

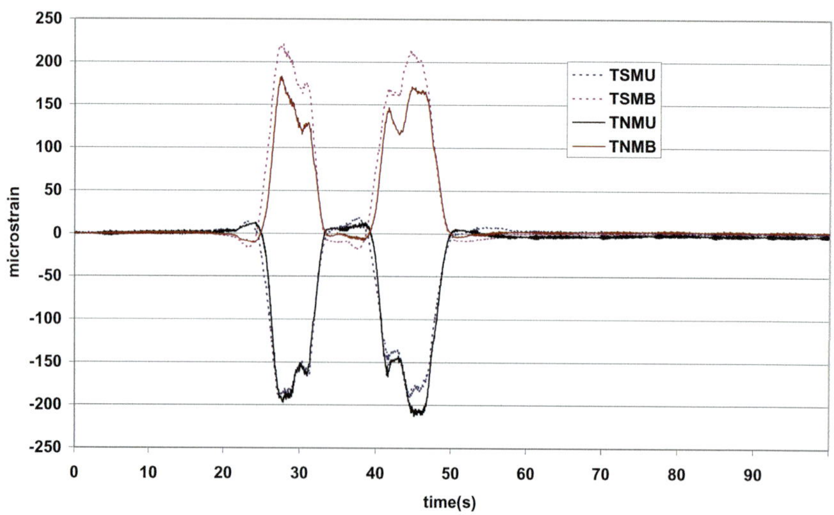

FIGURE 3.45 Strain in the upper and lower fibres at the midpoint of the northern and southern longitudinal beams during the passage of a single diesel locomotive.

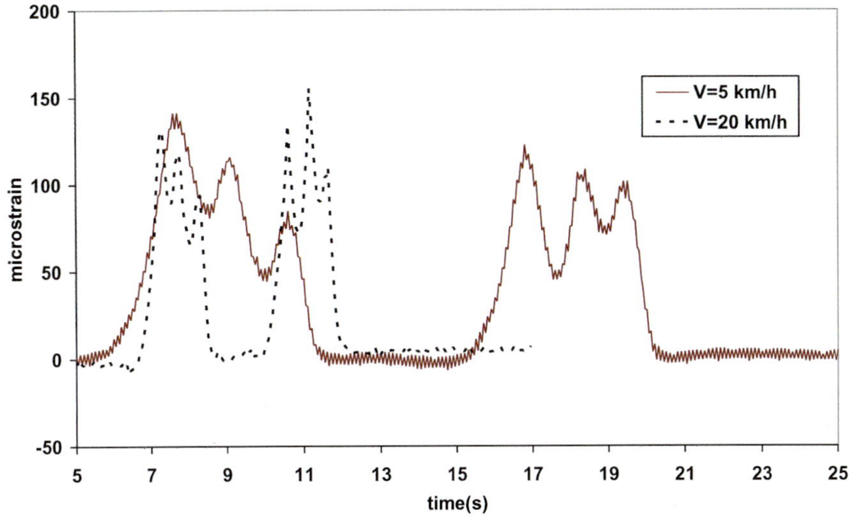

FIGURE 3.46 Strain in the upper and lower fibres at the midpoint of the northern and southern longitudinal beams during the passage of a single diesel locomotive at speeds of 5 and 20 km/h.

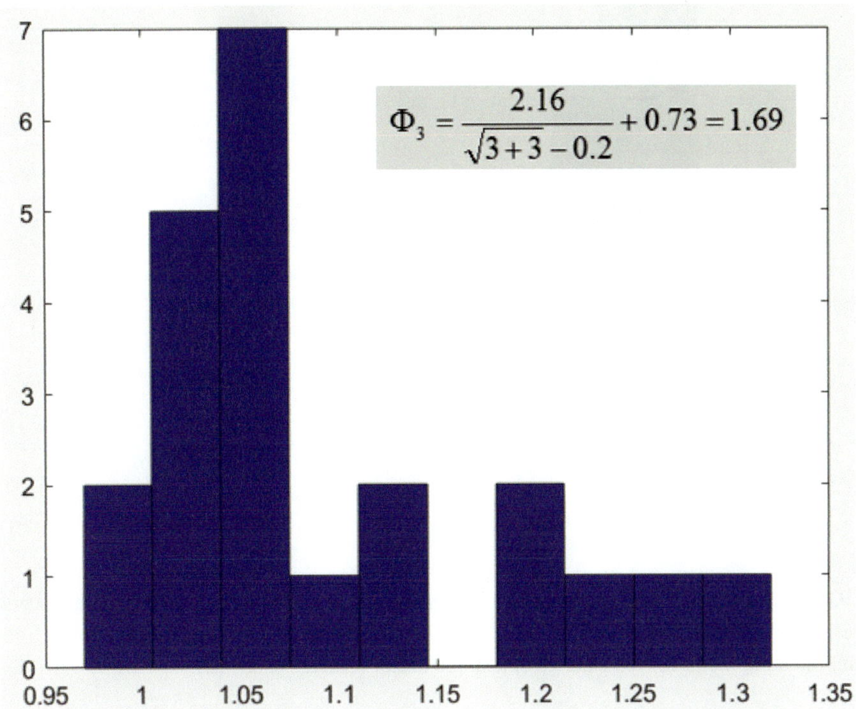

FIGURE 3.47 Histogram of the dynamic amplification factor (DAF) for strain at the mid-span of the longitudinal beam during the passage of a single diesel locomotive.

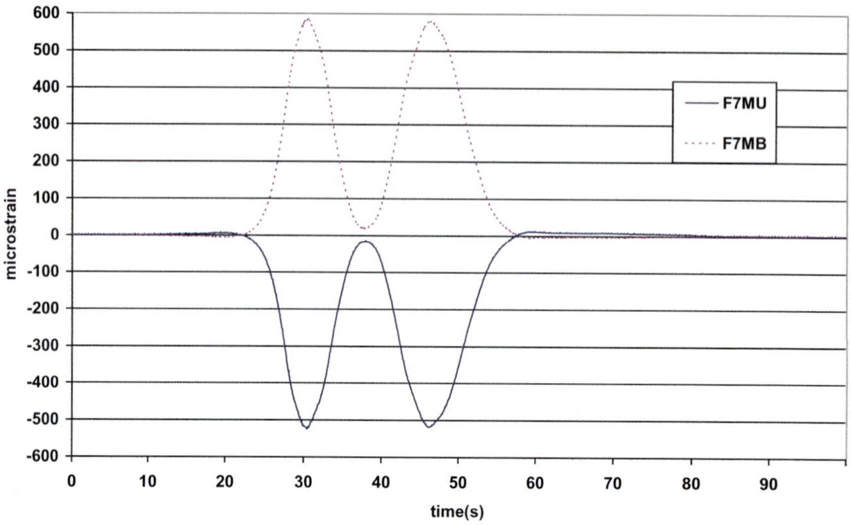

(a)

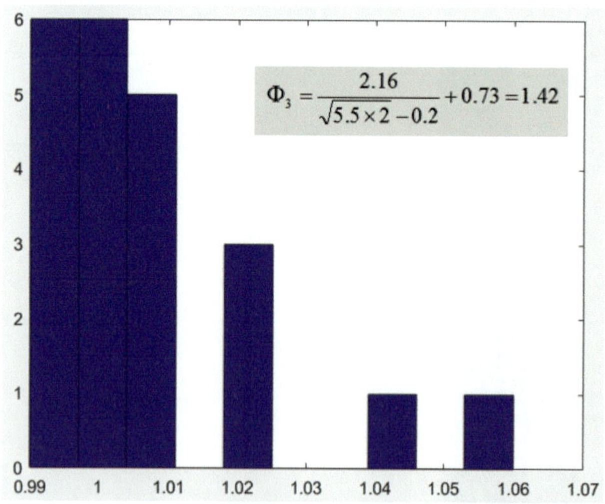

(b)

FIGURE 3.48 (a) Strain at the top and bottom fibre at the mid-span of the seventh transverse beam during the passage of a single diesel vehicle and (b) histogram of the dynamic amplification factor for the passage of a single diesel vehicle.

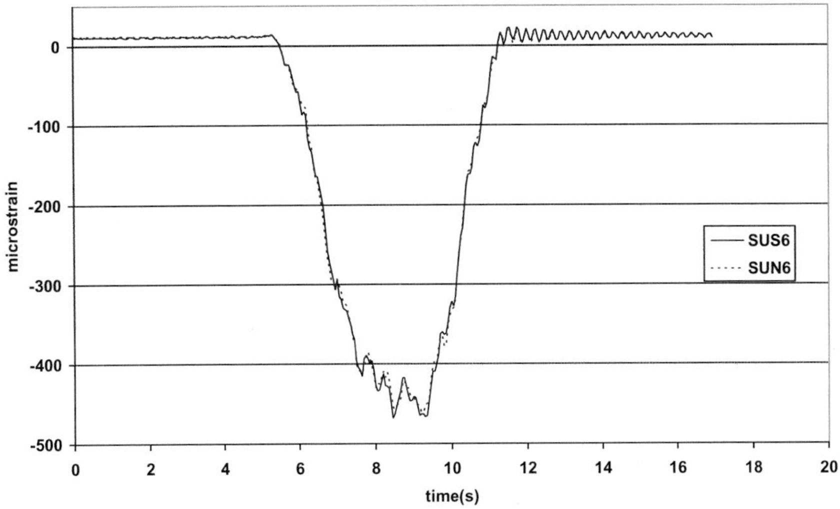

(a)

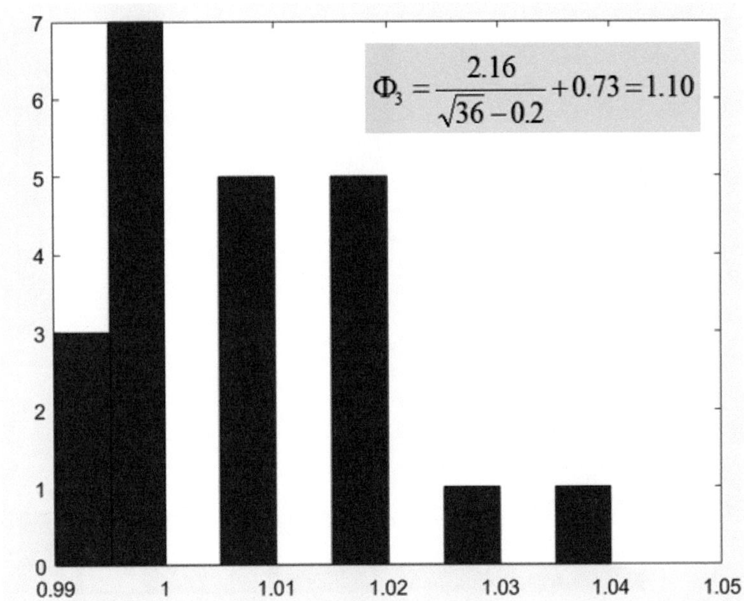

(b)

FIGURE 3.49 (a) Strain in the upper chord at the sixth panel point of the southern and northern trusses during the passage of a diesel locomotive and two wagons moving to the right at a speed of 50 km/h and (b) histogram of the dynamic amplification factor.

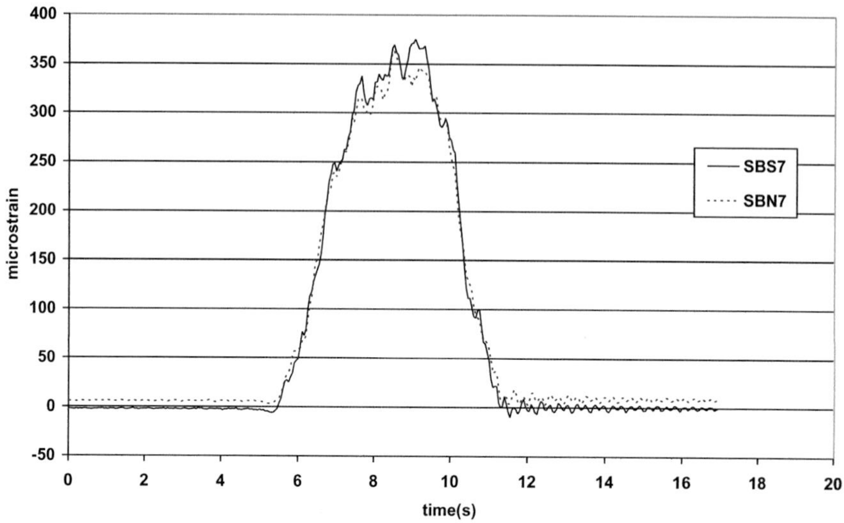

(a)

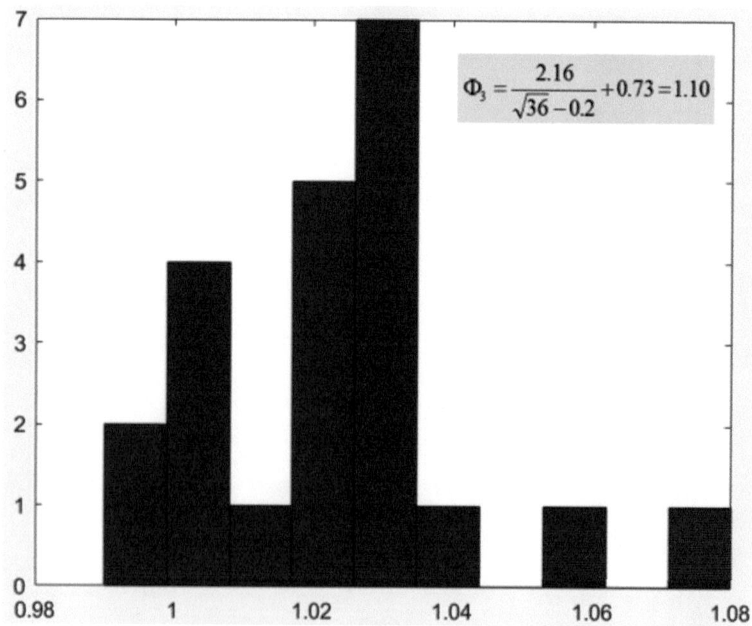

(b)

FIGURE 3.50 (a) Strain in the bottom chord at the seventh panel point of the southern and northern trusses during the passage of a diesel locomotive and two wagons moving from left to right at a speed of 50 km/h and (b) histogram of the dynamic amplification factor.

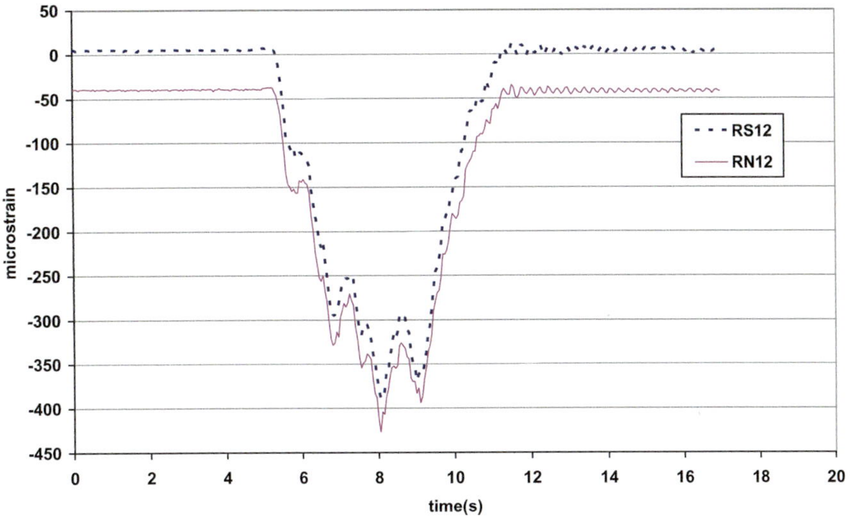

FIGURE 3.51 Strain in the lower chord at the 12th panel point of the southern and northern trusses during the passage of a diesel locomotive and two wagons moving from left to right at a speed of 50 km/h.

3.11.3 Comparison of Results with Analytical Models

The measured values obtained during the load testing can be compared to the analytical values derived from truss analysis. In the following sections, the truss will be analysed using the virtual work method and matrix analysis, with the results compared to those from the load testing. This comparison will help illustrate how the experimental data aligns with the analytical predictions.

3.11.3.1 Virtual Work Method

The virtual work method is a technique in structural analysis that allows us to calculate displacements in structures by applying a hypothetical or "virtual" unit load at the point where we want to measure the displacement. By analysing the internal forces in each truss member due to both the real load and the virtual load, we can determine the displacement at the chosen location through energy principles.

In this example, we want to find the vertical displacement at the mid-span of Neka Bridge for static positioning of a single locomotive at position 4, as shown in Figure 3.52a.

To calculate mid-span displacement, the following steps must be taken:

1. *Analyse the truss under the applied train load in tests*
   First, we analyse the truss under the actual loading condition (the train load). Using the method of joints, we calculate the internal force in each truss

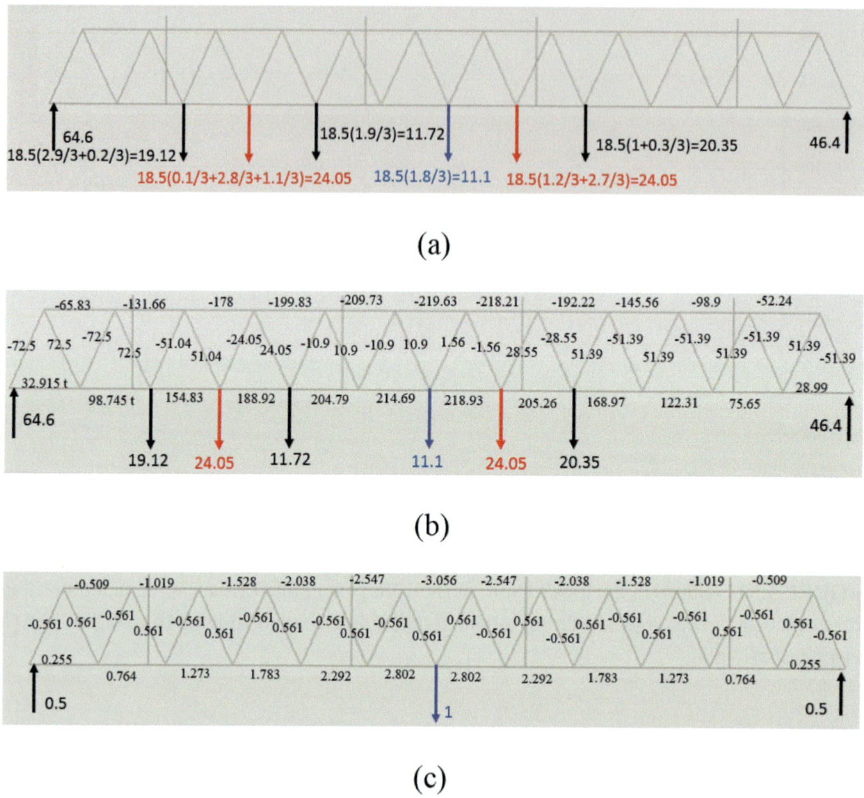

FIGURE 3.52 (a) Loading of the truss with a single diesel at a position 12 m from the right support (Position 4), (b) Truss analysis for the applied loading, and (c) Truss analysis under the unit vertical load of virtual work method at the mid-span.

member due to this load. This involves solving equilibrium equations at each joint to find the forces, which are shown in Figure 3.52b.

2. *Apply a virtual unit load at the mid-span*
   To calculate the displacement at the mid-span, apply a virtual unit load at this location in the direction of the desired displacement (in this case, vertically downward). Using the method of joints again, calculate the internal forces in each member due to this virtual unit load. These forces are shown in Figure 3.52c.
3. *Calculate displacement using the virtual work formula*
   Now, we can apply the virtual work formula to find the mid-span displacement:

$$\Delta_{min} = \sum \frac{N_i \times n_i \times l_i}{E_i \times A_i} \tag{3.10}$$

where

- N_i is the force in member i from the applied train load during tests,
- n_i is the force in member i from the virtual unit load,
- l_i is the length of member i,
- E_i is the modulus of elasticity of member i,
- A_i is the cross-sectional area of member i.

This formula essentially sums the contributions of each truss member to the displacement at the mid-span, taking into account both the real and virtual forces. By substituting the calculated forces N_i and n_i from Figure 3.52b and 3.52c, as well as the material and geometric properties l_i, E_i, and A_i, for each member into the formula, vertical displacement of 40 mm is achieved, which is slightly higher than 34 mm observed during load tests.

$$\Delta mid = 0.561$$

$$\times \frac{\left[4(72.5)+2(51.04)+2(24.05)+4(10.9)+2(1.56)+2(28.55)+8(51.39)\right]\times 3.3}{2\times 64.3\times 2000}$$

$$+[65.83\times 0.509+131.66\times 1.019+178\times 1.528+199.83\times 2.038$$
$$+209.73\times 2.547+219.63\times 3.056+218.21\times 2.547+192.22\times 2.038$$
$$+145.56\times 1.528+98.9\times 1.019+52.25\times 0.509+32.915\times 0.255$$
$$+98.745\times 0.764+154.83\times 1.273+188.92\times 1.783+204.79\times 2.292$$
$$+214.69\times 2.802+218.93\times 2.802+205.26\times 2.292+168.97\times 1.783$$
$$+122.31\times 1.273+75.65\times 0.764+28.99\times 0.255]\times 3/(2\times 140\times 2000)$$

$$= 0.04\,\text{m}.$$

3.11.3.2 Matrix Analysis Method

With a two-dimensional modelling of the truss, matrix analysis of the structure can be performed by forming the stiffness matrix. Each node in the truss has two degrees of freedom (horizontal and vertical displacements). For a truss with 25 nodes, this results in a total of 50 degrees of freedom, as shown in Figure 3.53.

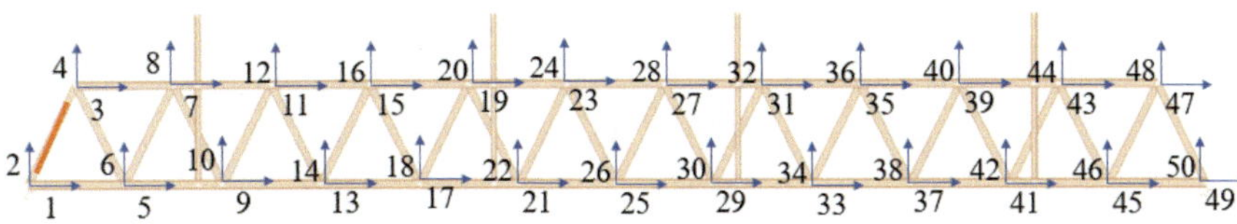

FIGURE 3.53 Degrees of freedom for the truss (two degrees of freedom per node, totalling 50 degrees of freedom).

Each element in the truss (modelled as an axial element) has a stiffness matrix based on its orientation and properties. the stiffness matrix for an element inclined at an angle θ is:

$$\frac{EA}{L}\begin{bmatrix} C2 & CS & -C2 & -CS \\ CS & S2 & -CS & -S2 \\ -C2 & -CS & C2 & CS \\ -CS & -S2 & CS & S2 \end{bmatrix} \tag{3.11}$$

Here, $\frac{EA}{L}$ is a scaling factor, where E is the modulus of elasticity, A is the cross-sectional area, and L is the length of the element. C2, CS, and S2 are shorthand for trigonometric expressions based on the angle of the element:

- $C2 = (\cos\theta)^2$ (3.12)
- $CS = (\cos\theta)(\sin\theta)$ (3.13)
- $S2 = (\sin\theta)^2$ (3.14)

In Neka Bridge, elements have two different angle values θ:

1. 63° (diagonal members)
2. 0° (Horizontal members)

For example, for diagonal elements,

- $C2 = (\cos\theta)^2 = 0.454^2 = 0.206$
- $CS = (\cos\theta)(\sin\theta) = 0.454 \times 0.891 = 0.404$
- $S2 = (\sin\theta)^2 = 0.891^2 = 0.794$

Thus, the stiffness matrix for a diagonal element is:

$$\frac{2\times10^6\times64.3}{303}\begin{bmatrix} 0.206 & 0.405 & -0.206 & -0.405 \\ 0.405 & 0.794 & -0.405 & -0.794 \\ -0.206 & -0.405 & 0.206 & 0.405 \\ -0.405 & -0.794 & 0.405 & 0.794 \end{bmatrix}$$

The individual stiffness matrices for each element are assembled into a global stiffness matrix for the entire structure. This matrix, which encompasses all degrees of freedom in the truss, allows us to analyse how the entire truss structure will respond to applied forces.

Next, boundary conditions are defined, and global stiffness matrix is modified to account for supports and constraints. For example, if a node is fixed in both directions, the corresponding rows and columns in the global stiffness matrix are removed to reflect that it has known displacement. Assuming a pinned support at the left end and a roller support at the right end, the first, second, and fiftieth rows and columns of the stiffness matrix are removed, resulting in a symmetric square matrix with dimensions of 47×47.

Now it is possible to create a load vector representing the external forces applied to each degree of freedom (e.g., forces at specific nodes). This vector should be of the same dimension as the modified global stiffness matrix. For the load formation presented above in Figure 3.52a, a vector with single concentrated load at nodes 10, 1, 18, 26, 30, and 34 is created, details of which are given in Table 3.3.

It is now possible to calculate displacement at different nodes of the truss. To do so, the below system of linear equations is solved,

$$[K]\{d\}=\{F\} \tag{3.15}$$

where:

- $[K]$ is the global stiffness matrix (after applying boundary conditions),
- $\{d\}$ is the vector of nodal displacements,
- $\{F\}$ is the vector of external forces.

This solution gives the displacement of each node in the truss. The displacements at several nodes are presented in Table 3.4. At node 26 (middle of the span), the vertical displacement is calculated to be 43 mm.

TABLE 3.3
DOF of Neka Bridge and applied load under loading of Figure 3.52a

DOF	10	14	18	26	30	34
Force (ton)	−20.35/2	−24.05/2	−11.1/2	−11.72/2	−24.05/2	−19.12/2

TABLE 3.4
Vertical displacement at several nodes of Neka truss bridge from matrix analysis method

DOF	6	10	14	18	22	26
Displacement(cm)	−1.1	−2.2	−3.1	−3.8	−4.2	−4.3
DOF	30	34	38	42	46	–
Displacement(cm)	−4.1	−3.7	−3.0	−2.1	−1.1	

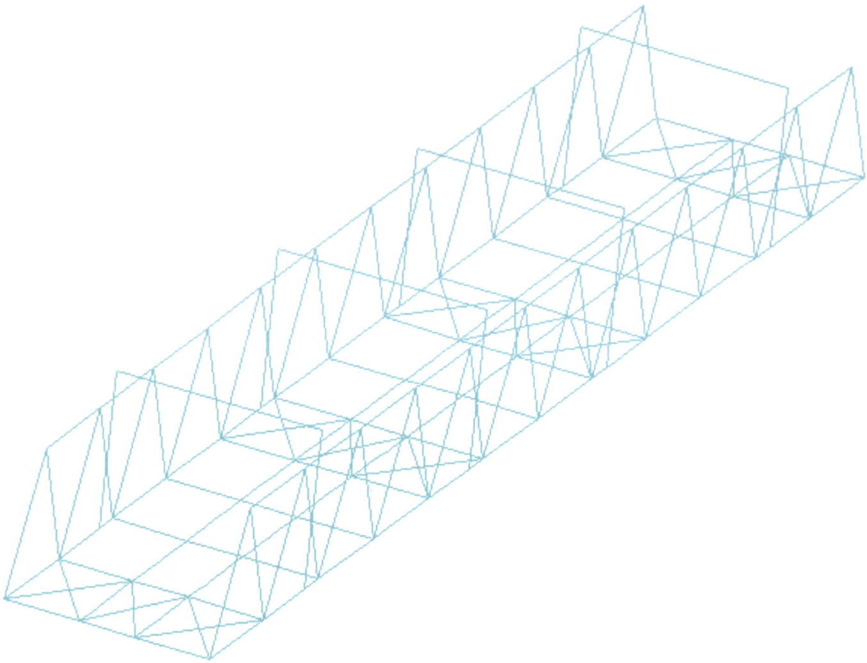

FIGURE 3.54 3D finite element model with ANSYS software.

It is evident that this method produces values that are approximately 26% higher than those observed during the field test. Hence, it is evident that performing test has the potential to give more realistic values.

3.11.3.3 FE Model

It is also possible to develop a detailed 3D model of the bridge using a finite element analysis (FEA) software, as shown in Figure 3.54. Developing such a model is considerably more complex and time-consuming, requiring a higher level of detail, but it provides more accurate results compared to other methods. Additionally, this approach allows for the analysis of the structure under dynamic loads, making it superior for certain applications. Figure 3.55 presents a comparison of displacements at various points on the Neka Bridge from the FE model and experimental load tests. As shown, the results are nearly identical, demonstrating the accuracy of the FE model.

3.12 SUMMARY

In this chapter, we explored the various advanced instrumentation tools used in condition monitoring, with a particular focus on sensors that aid in the assessment of structural integrity. We began by classifying global monitoring sensors, which provide an overarching view of structural health, followed by an in-depth discussion on

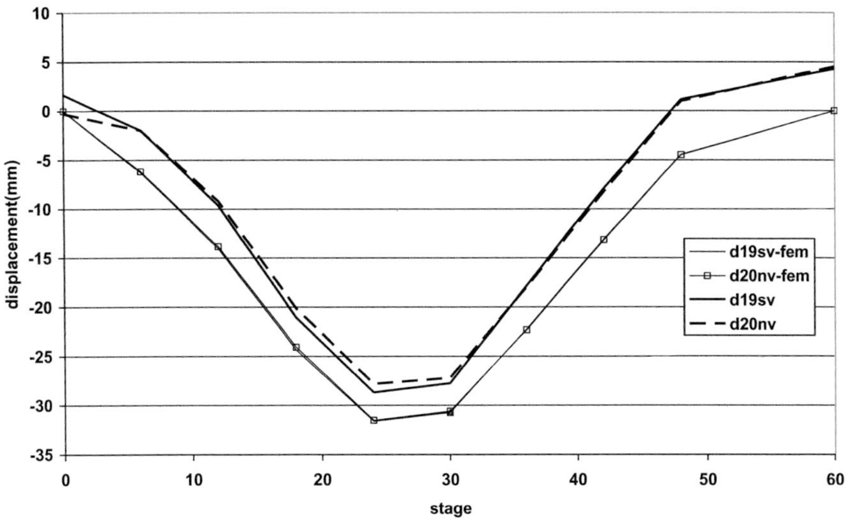

FIGURE 3.55 Comparison of vertical displacement measured under static load from single diesel and finite element model at 19 m of the bridge.

strain measurement sensors. Through a case study on monitoring strain in a railway bridge, we illustrated the practical application of these tools in real-world scenarios. Strain measurement is critical for understanding the behaviour of structural elements.

We then examined displacement measurement sensors, which are vital for detecting relative movements between structural components, such as between masonry blocks under load. This is particularly important for assessing the stability of ageing structures under dynamic loads like those encountered in railway bridges.

Next, we discussed vibration measurement sensors, using the Tehran Metro turnout as an example. Vibration monitoring is a key component in assessing the condition of structural components subjected to dynamic loads, ensuring the longevity and safety of the system.

Finally, we introduced data acquisition systems, which serve as the backbone of monitoring setups, allowing for the continuous collection and analysis of sensor data.

After explaining the different types of sensors, the chapter introduced the process of raw data processing, which is crucial for eliminating any unwanted noise introduced during experimental testing.

The chapter concluded with a case study on the load testing of the Neka Bridge, demonstrating how the concepts and methods discussed can be applied in real-world scenarios. The results presented will be used in the next chapter to assess the load rating factor of the bridge.

This chapter laid the groundwork for understanding how these tools are used to monitor critical parameters in structures under load. In the next chapter, we will shift our focus to the assessment of load-carrying capacity. By integrating the sensor data discussed in this chapter with theoretical models, we will evaluate the bridge's ability to safely accommodate increased loads.

DISCUSSION QUESTIONS

1. What are the differences in the capacity of the accelerometer used in the bridge, the sleeper, the axle box of the wagon, the central piece of the railway turnout, and the ground surface?
2. If the passing load is as shown in Figure 3.12, what is the estimated passing speed in the monitoring systems shown in Figure 3.13?
3. How can the bridge's measured strain in Figures 3.45 and 3.46 be interpreted? Based on interpretation, what decisions can be made?

REFERENCES

[1] J. F. Lin, X. Y. Li, J. Wang, L. X. Wang, X. X. Hu, and J. X. Liu, "Study of Building Safety Monitoring by Using Cost-Effective MEMS Accelerometers for Rapid After-Earthquake Assessment with Missing Data," *Sensors*, vol. 21, no. 21, pp. 7327, 2021.

[2] J. W. Baldwin, H. Salane, and R. Duffield, "Fatigue Cracks and Their Detection," *Transportation Research Record.*

[3] R. Pullin, K. Holford, J. Robert, and M. Eaton, "Acoustic Emission Monitoring of Bridge Structures in the Field and Laboratory," *Journal of Acoustic Emission*, vol. 26, pp. 172–181, 2008.

[4] X. Zhao et al., "Bridge Displacement Monitoring Method Based on Laser Projection-Sensing Technology," *Sensors (Basel)*, vol. 15, no. 4, pp. 8444–8463, 2015.

[5] L. Miccinesi et al., "Structural Health Monitoring of the Maratona Tower at Artemio Franchi Stadium," *Structural Monitoring and Maintenance*, 2019.

[6] T. Chan, "Structural Monitoring of Tsing Ma Bridge Using Fiber Bragg Grating Sensors," *IEEE*, 2004.

[7] B. Zheng, "Research on Structural Deformation Monitoring Based on Digital Close-Range Photogrammetry Technology," 2024, [Online]. Available: www.atlantis-press.com/proceedings/iccase-24/126001224

[8] "Publication 139: Bridge Loading Code." https://shaghool.ir/downloadarea.php?id=1590

[9] J. Brownjohn, "Structural Health Monitoring of Civil Infrastructure," *Philosophical Transactions A: Mathematical, Physical and Engineering Sciences*, 2016. DOI: 10.1098/rsta.2006.1925

[10] S. Cho, S. Jang, H. Jo, and K. Mechitov, "Structural Health Monitoring System of a Cable-stayed Bridge Using a Dense Array of Scalable Smart Sensor Network," in *Proceedings of SPIE—The International Society for Optical Engineering*, 2010.

[11] "ISO 16063.21:2003 Methods for the Calibration of Vibration and Shock Transducers," 2024.

[12] A. Rytter, *Vibrational Based Inspection of Civil Engineering Structures*. Aalborg University Press, 1993.

[13] M. Stéphane, *A Wavelet Tour of Signal Processing*, 3rd ed. Academic Press, 2009.

[14] K. Steiglitz, *A Digital Signal Processing Primer*. Prentice Hall, 1996.

[15] A. V. Oppenheim, R. W. Schafer, and J. R. Buck, *Discrete-Time Signal Processing*, 3rd ed. Pearson Prentice Hall, 2010.

4 Load Rating

4.1 INTRODUCTION

Bridge load rating assesses the structural capacity of a bridge member to ensure safety, serviceability, and durability. It involves determining the bridge's ability to handle specified loads, including the live design load, legal load, or loads specified for issuing passage permits. Various structural evaluation methods align with structural design approaches such as the allowable stress method, strength method, and load and resistance factor rating (LRFR). Each method has its own set of calculations, load factors, and safety margins, applied based on the bridge's age, design, and the desired level of conservatism. This chapter will elaborate the LRFR, which is currently one of the most prevalent approaches in structural assessment. To enhance understanding, the theoretical concepts will be illustrated through a practical case study, demonstrating how the method is systematically applied in real-world engineering scenarios.

4.2 LOAD RATING FACTOR (RF) CALCULATION

The load rating factor (RF) measures the bridge member's capacity relative to its load demand. It is determined according to the following general equation:

$$\mathrm{RF} = \frac{C - A_1 D}{A_2 L(1+I)} \tag{4.1}$$

where

C is the capacity of the bridge member.
A_1 the dead load factor.
D the dead load.
A_2 the live load factor.
L the live load.
I the dynamic allowance factor.

DOI: 10.1201/9781003494652-4

Each evaluation method has unique calculation steps and equations to determine RF based on specific load factors and safety margins.

4.2.1 Load and Resistance Factor Rating

LRFR is the most advanced and widely adopted approach for calculating load ratings. It is a comprehensive method that differentiates between extreme loading conditions (ultimate limit state (ULS)) and routine, operational conditions (serviceability limit state (SLS)). This dual focus ensures that the bridge can safely handle both peak demands and regular usage without compromising its structural integrity or user comfort.

ULS is used to assess the bridge's resilience under "worst-case" loading conditions, such as heavy traffic or severe weather. This limit state considers the maximum possible stress the bridge might face and applies conservative load factors to ensure it does not experience catastrophic failure under these conditions. The ULS formula for RF in LRFR at inventory level is:

$$\mathrm{RF} = \frac{C - 1.25\,\mathrm{DC} - 1.5\,\mathrm{DW}}{\alpha L(1+I)} \tag{4.2}$$

where

- DC represents the dead load from the bridge's structural components.
- DW represents dead load of wearing surfaces and utilities.
- A dead load factor of 1.25 and an additional factor of 1.5 for DW ensure that any uncertainties in these weights are accounted for.
- α is the live load factor, which is set to 1.75 for inventory level and 1.35 for operating level, reflecting the increased demand expected under extreme scenarios.

The ULS calculation is ideal for evaluating the bridge's load-bearing capacity during unusual events, such as heavy truck traffic or storms. This layer of conservatism helps ensure that the bridge will remain safe and functional during these times, minimizing the risk of sudden failure and enhancing the safety of users.

LRFR is the preferred method in modern bridge engineering because it offers a more nuanced approach, addressing both extreme and normal conditions with tailored safety factors. This method's flexibility and comprehensiveness make it the best option for achieving a balance between maximum safety and practical usability, complying with regulatory standards, and ensuring a long service life for the bridge.

4.2.2 Step-by-Step Load Rating Process

The load rating process includes several key steps to determine a bridge member's load-carrying capacity accurately:

1. Calculate member capacity (C):
   - Represents the maximum load-bearing capacity of a bridge member based on material strength, design, and cross-sectional area.
2. Assess dead and live loads:
   - Dead load (D): Constant weight from the bridge's structural components.
   - Live load (L): Variable loads from vehicles, pedestrians, and environmental factors, fluctuating with usage patterns.
3. Apply the dynamic allowance factor (I):
   - Adjusts the live load to account for dynamic effects like moving vehicles and vibrations. Calculated as $L(1+I)$, this factor ensures that both static and dynamic stresses are covered.
4. Subtract dead load demand from capacity:
   - Calculates the remaining capacity for live loads: $C–A_1·D$. This step isolates capacity for live loads by accounting for the bridge's self-weight.
5. Calculate the load-rating factor (RF):
   - Divides the remaining capacity by the live load demand ($\alpha L(1+I)$) to measure the member's ability to support live loads adjusted for dynamics:

$$\mathrm{RF} = \frac{\text{Remaining Capacity for Live Load}}{\text{Live Load Demand (adjusted for dynamics)}} \tag{4.3}$$

6. Interpret the load-rating factor:
   - If $\mathrm{RF} \geq 1$: The member has sufficient capacity to support the assessed load.
   - If $\mathrm{RF} < 1$: The member lacks adequate capacity, necessitating reinforcement, load limits, or additional safety measures.

To calculate the load-rating factor (RF) for a bridge, several key variables must be determined: C, L, I, and D. The variable D, representing the dead load, is based on the current weight and specific elements of the bridge, and doesn't require a particular design standard. However, C (capacity), L (live load), and I (dynamic allowance factor) and the dead and live load factors A_1 and A_2 vary depending on the level at which the load rating assessment is conducted. The flowchart in Figure 4.1 shows an overview of the staged assessment process according to *AASHTO* [3].

Each stage progressively refines the assessment by making more accurate assumptions regarding the loading conditions and bridge characteristics. The process begins with basic, conservative assumptions about the loads and the bridge's capacity. If the bridge meets the required rating factor (RF) at this initial level, then it is deemed adequate for its intended use, and no further analysis is required.

However, if the bridge does not pass at this initial stage, the assessment continues to the next level, where the load assumptions and bridge characteristics are adjusted to reflect more realistic conditions. Sometimes, a bridge may not pass under these standard, conservative assumptions, but as the evaluation progresses to the next stages with more detailed adjustments, it might meet the necessary criteria.

The final stage, which involves detailed analysis and potentially load testing, is the most precise. It incorporates any "hidden" capacity the bridge may have that isn't captured

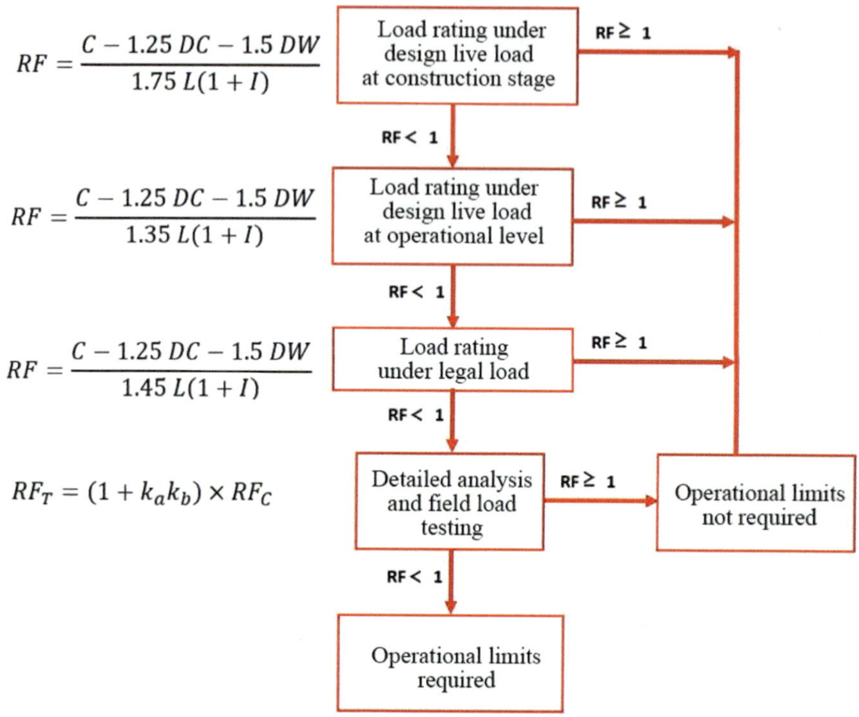

FIGURE 4.1 Stages of load rating factor assessment in accordance with AASHTO LRFD.

in the earlier conservative assumptions. If the bridge still does not satisfy the load rating requirements at this stage, then operational restrictions or reinforcement measures will be recommended to ensure safety. This step-by-step, staged approach allows for an efficient and thorough evaluation, balancing simplicity in initial assumptions with accuracy in later stages to avoid unnecessary restrictions when a bridge may indeed be adequate. In the following sections, details for calculating *C*, *L*, and *I* are discussed.

4.2.3 Calculation of Capacity (*C*)

The capacity is a critical component in determining the load rating factor, as it represents the maximum load the bridge member can safely support. The AASHTO LRFD code provides a framework for calculating *C*, based on a nominal capacity (R_n) and applying reduction factors to account for uncertainties in capacity calculation, load conditions, and the bridge's current state.
Key Reduction Factors:

1. *General reduction factor* (ϕ): Adjusts for uncertainties in capacity calculation. This factor ensures that the bridge can still perform safely even when variations in material strength or load assumptions occur.
2. *Condition factor* (ϕ_c): Reflects the current state of the bridge element, such as degradation or environmental effects. A higher value indicates better structural

TABLE 4.1
Condition factor for structural elements (MBE)

ϕ_c	Structural Condition
1	Good
0.95	Fair
0.85	Poor

TABLE 4.2
System factors for bending and axial behaviour (MBE)

ϕ_s	Bridge Type
0.85	Welded members in bridges with two girders/trusses/archs
0.9	Riveted members in bridges with two girders/trusses/archs
0.9	Members with multiple wide flanges in truss bridges
0.85	Bridges with three girders spaced 1.8 m apart
0.95	Bridges with four girders spaced less than 1.2 m apart
1	Other girder bridges and slab bridges
0.85	Cross girders with spacing over 4 m and non-continuous longitudinal beams
1	Subsystems with redundancy between cross girders

condition, while lower values indicate deterioration. Table 4.1 provides condition factors for structural elements.

3. *System factor* (ϕ_s): Accounts for the bridge element's behaviour within the overall bridge system, including redundancy and load-sharing between components. Elements that work well in a redundant system may have a higher ϕ_s factor. For example, welded members in bridges with two girders have a ϕ_s of 0.85, while bolted members in similar bridges have a ϕ_s of 0.9. Table 4.2 provides a list for different values of system factors.

The adjusted C is calculated by multiplying these factors with the nominal capacity (R_n):

$$C = \phi_c \cdot \phi_s \cdot \phi \cdot R_n \tag{4.4}$$

where the product of ϕ_c and ϕ_s should not be less than 0.85 for any scenario.

4.2.4 Calculation of Live Load (*L*) and Dynamic Allowance Factor (*I*)

The live load (*L*) for designing railway bridges is typically derived from standard loading patterns based on the specific axle loads on the track. In Iran, the standard live load for railway bridges is provided in Figure 4.2 from Publication 139 [2]. This load pattern applies to railway tracks with an axle load of 25 tons. If the axle load is different, the load pattern is adjusted accordingly as below.

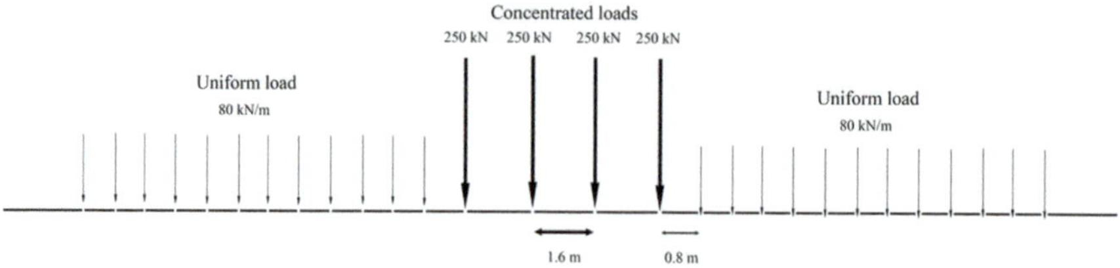

FIGURE 4.2 Design live load configuration from Publication 139 [2].

- For an axle load of 20 tons (referred to in Publication 139 as 'Light type 2'), the pattern is multiplied by 0.8.
- For an axle load of 30 tons (referred to in Publication 139 as 'Heavy type 3'), the pattern is multiplied by 1.2.

The dynamic allowance factor (*I*) is calculated according to the Iranian code [2], and it accounts for the increased load effects due to the dynamic nature of moving traffic. It should be highlighted that the dynamic allowance factor in the Iranian code is adapted from UIC 776 [4]. The equation for calculating the dynamic allowance factor is shown as Equation 4.5. The value of L_φ in this equation depends on the length of the member being analysed. For instance, in a single-span truss, L_φ is equal to the span length. For transverse and longitudinal deck beams in a truss, L_φ should be twice the length of the beam. Equation 4.5 for the dynamic allowance factor is:

$$\Phi_3 = \frac{2.16}{\sqrt{L_\Phi} - 0.2} + 0.73 \leq 2 \tag{4.5}$$

4.3 EVALUATION OF LOAD RF OF NEKA BRIDGE

In this example, we are analysing Neka truss bridge that was discussed in Chapter 3. The truss members have the following geometrical and material properties:

- Horizontal truss members (ST52) are IPBL240+2PL200×10 of 3 m in length, with the following parameters:

$$I_x = 9656\,\text{cm}^4$$

$$I_y = 9637\,\text{cm}^4$$

$$A = 140.3\,\text{cm}^2$$

$$r_x = r_y = 8.3\,\text{cm}$$

$$F_u = 5200 \, {}^{\text{kg}}\!/_{\text{cm}^2}$$

$$F_y = 3600 \, {}^{\text{kg}}\!/_{\text{cm}^2}$$

- Diagonal truss members (ST52): The diagonal members are (IPBL220) with a length of 3.3 m and at a 63° angle, as shown in Figure 4.3.

$$I_x = 5410\,\text{cm}^4$$

$$I_y = 1950\,\text{cm}^4$$

$$A = 64.3\,\text{cm}^2$$

$$r_x = 9.17$$

$$r_y = 5.51\,\text{cm}$$

$$F_u = 5200 \, {}^{\text{kg}}\!/_{\text{cm}^2}$$

$$F_y = 3600 \, {}^{\text{kg}}\!/_{\text{cm}^2}$$

- Cross beams (ST52): Cross beams (IPBL400) have a section area of 159 cm^2 and a section modulus (S) of 2310 cm^3.
- Longitudinal beams (ST52): Longitudinal beams (IPBL240) have a section area of 76.8 cm^2 and a section modulus (S) of 675 cm^3.

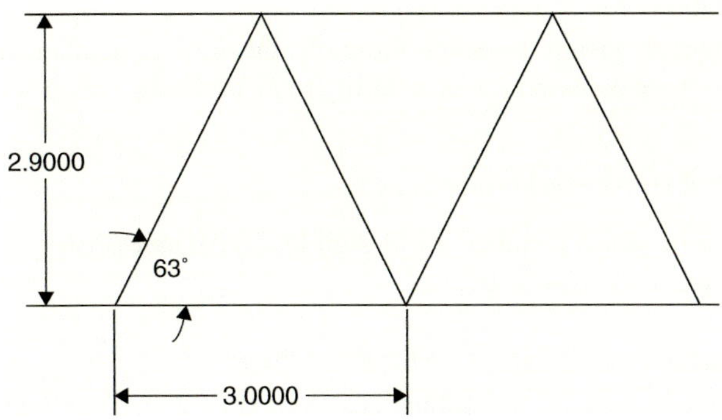

FIGURE 4.3 Details of diagonal members of Neka Bridge.

The bridge carries the load of a railway line with an axle load of 20 tons. The overall load placed on the bridge is calculated to be 2694 kN, based on the light-weight live load pattern as specified by the Iranian design code. The dynamic load factor for the truss members is calculated, considering the full length of the truss, cross beams, and longitudinal beams, resulting in different dynamic factors for each part of the truss.

To evaluate the load-bearing capacity of the truss, we need to analyse the following aspects of the members, following the stages presented in the flowchart in Figure 4.1.

1. Flexural evaluation of the members (bending capacity)
   - *Longitudinal beams*: These beams are subject to bending moments, and we calculate the maximum bending moment due to both dead load and live load. The bending demand and resistance are evaluated to determine if the beam can resist the applied loads without failure. This includes calculating the fatigue resistance, flexural capacity, and shear capacity.
   - *Cross beams*: Similar calculations are done for the cross beams, including the assessment of bending, shear, and fatigue resistance under the same loading conditions.
2. Axial load evaluation (member in tension or compression)
   - Lower chord (tensile member): We assess the axial load in the upper chord members of the truss to ensure that they can resist the applied tensile forces without failure. The lower chord in the middle of the span is the member with the maximum tension, and if it is okay, it can be said that all horizontal truss elements in tension are fine.
   - Upper chord (compressive member): For the lower chord members, we evaluate their compression capacity to ensure they can resist buckling or crushing under the applied loads. upper chord in the middle of the span is the element with the maximum compression, and if deemed fine, it can be concluded that all horizontal truss elements in compression are fine.
   - Diagonal compression members: These members, which are subject to axial compressive forces, are evaluated for buckling resistance and overall stability.
   - Diagonal tension members: Similarly, the tension members in the diagonal braces are assessed for their ability to resist tensile forces without failure.

4.3.1 Load Rating for Design Load

The first step is to assess members for design loads. An assessment of the inventory level is presented in the following.

4.3.1.1 Flexural Assessment: Longitudinal Beams

Neka bridge has 12 sets of longitudinal beams of IPBL240, each with a length of 3 m. Let's assume the dead load is 14 kN/m. The critical positioning of design live load (shown in Figure 4.2) is presented in Figure 4.4.

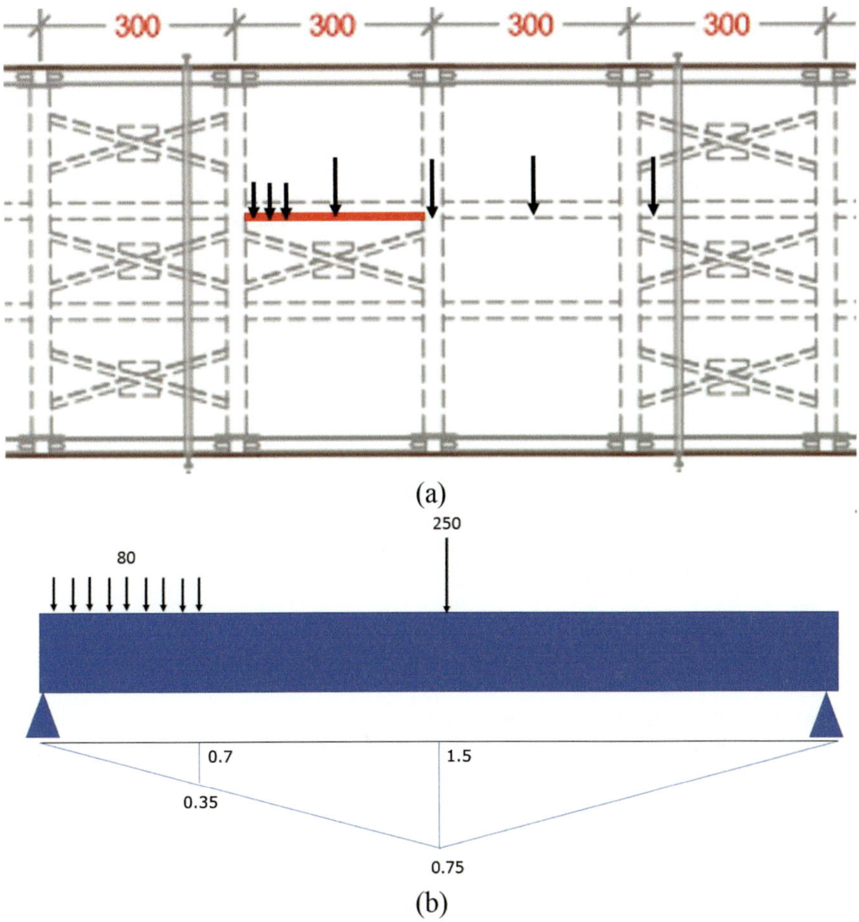

FIGURE 4.4 Critical position of live load on the longitudinal beam: (a) plan view and (b) side view (influence line).

First, we calculate the reactions at the supports due to both the point load and the UDL. Let

- R_A = Reaction at the left support.
- R_B = Reaction at the right support.

Using the principle of static equilibrium for sum of vertical forces and sum of moments about Point A by taking moments about point A to solve for R_B:

$$R_A + R_B = 56 + 250 = 306\,\text{kN}$$

$$R_B \times 3 = (56 \times 0.35) + (250 \times 1.5) = 19.6 + 375 = 394.6$$

$$R_B = \frac{394.6}{3} = 131.53\,\text{kN}$$

Now, using $R_A + R_B = 306\,\text{kN}$:

$$R_A = 306 - 131.53 = 174.47\,\text{kN}$$

The bending moment at the centre, where the point load is applied, will be the maximum moment. We can calculate it by considering the contributions from the reaction at A, the equivalent load from the UDL, and the point load, as shown below.

$$M_{\text{max}} = (R_A \times 1.5) - (56 \times (1.5 - 0.35)) = 197.3\,\text{kNm}$$

This can also be calculated using the influence line presented in Figure 4.4.b. For the concentrated load, we have:

$$M_{\text{point}} = 250 \times 0.75 = 187.5\,\text{kN.m}$$

For the uniform load, the bending moment can be calculated by integrating the influence line times the distributed load, as shown below

$$M_{\text{UDL}} = \int_0^{0.7} 80.f(x)\,\text{d}x$$

In which $f(x)$ is the influence function based on its slope (0.75/1.5=0.5). therefore:

$$M_{\text{UDL}} = \int_0^{0.7} 80.(0.5x)\,\text{d}x$$

$$M_{\text{UDL}} = 40 \times \left[\frac{x^2}{2}\right]^{0.7} = 40 \times \frac{(0.7)^2}{2} = 9.8\,\text{kN/m}$$

Therefore, the total bending moment at the centre of the beam is:

$$M_{\text{max}} = M_{\text{point}} + M_{\text{UDL}} = 187.5 + 9.8 = 197.3\,\text{kN/m}$$

Now it is possible to perform load rating factor for longitudinal beam for fatigue resistance, flexural capacity, and shear capacity.

4.3.1.1.1 Fatigue Resistance

Fatigue resistance is a critical aspect of structural design, particularly for steel components subjected to repeated loading cycles, such as bridges. According to AASHTO LRFD 6.6.1, the evaluation of fatigue resistance involves calculating the demand and resistance against fatigue to ensure that structures can withstand the anticipated number of load cycles without initiating and developing cracks or experiencing failure. The process includes determining the factored moment due to live load and impact, calculating the equivalent stress range, and comparing the induced stresses to the material's fatigue resistance.

First, the factored moment due to live load and impact is calculated as shown below:

$$M_{\text{LL+IM}} = 0.5 \times 0.8 \times M_{\text{base}} \times \Phi_3 \tag{4.6}$$

where M_{base} is the maximum bending moment under live load, and Φ_3 is the dynamic load factor (Equation 4.6), which is determined as shown below:

$$\Phi_3 = \frac{2.16}{\sqrt{3+3} - 0.2} + 0.73 = 1.69$$

Therefore,

$$M_{\text{LL+IM}} = 0.5 \times 0.8 \times 197.3 \times 1.69 = 133\,\text{kN.m}$$

Next, we need to determine the nominal fatigue stress range, $(\Delta F)_n$ using the following formula:

$$(\Delta F)_n = \left(\frac{K}{N}\right)^{1/3} \tag{4.7}$$

where K is material constant related to fatigue detail category, and N is equivalent number of load cycles. For N, it is assumed that bridge has a service life of 75 years, and each day, 40 trains each with 10 wagons (with two bogies) cross over it. Therefore:

$$N = 365 \times 75 \times 2 \times (40 \times 10) = 21{,}900{,}000$$

And considering that the element is rolled steel, it is fatigue category A, with a K value of 82×10^{11} [1]. Therefore:

$$(\Delta F)_n = \left(\frac{82 \times 10^{11}}{21{,}900{,}000}\right)^{\frac{1}{3}} = 72.1\,\text{MPa}$$

Fatigue II load combination

$$0.8 \times M_{\text{LL+IM(mid)}} = 0.8 \times 133 = 106.4 \text{ kN.m}$$

Not OK

$$\sigma = \frac{M}{S} = \frac{106.4 \times 10^6}{675000} = 157.6 < 72.1$$

$$\text{RF} = \frac{72.1}{157.6} = 0.46 < 1$$

Fatigue I load combination

$$1.75 \times M_{\text{LL+IM(mid)}} = 1.75 \times 133 = 232.75 \text{ kN.m}$$

Not OK

$$\sigma = \frac{M}{S} = \frac{232.75 \times 10^6}{675000} = 345 < (\Delta F)_{TH} = 168$$

$$\text{RF} = \frac{168}{345} = 0.49 < 1$$

It is evident that longitudinal beams fail the design load in fatigue capacity, given their RF is under 1.

4.3.1.1.2 Flexural Capacity

Calculation of demand and resistance for flexural capacity is done in accordance with ASHTO LRFD 6.10.8 [1], which outlines the procedure for determining the flexural resistance of steel I-sections. The process involves several critical checks:

- Evaluation of the web load-shedding factor (R_b) to account for potential web buckling
- Assessment of local buckling resistance for the compression flange
- Calculation of lateral–torsional buckling resistance
- Determination of the flexural resistance as the minimum of local and lateral–torsional buckling resistances
- Stress control verification to ensure the factored moment does not exceed the factored flexural resistance; and
- Computation of the rating factor (RF) to quantify the girder's capacity relative to applied loads.

These checks collectively ensure the structural integrity of steel I-girders under various failure modes, providing a comprehensive evaluation of flexural capacity for both design and assessment purposes.

4.3.1.1.3 Web Load-Shedding Factor (R_b)

The web load-shedding factor R_b accounts for the reduction in flexural resistance due to web buckling in slender-web sections. The steps to check are as follows:

1. Calculate the web compactness ratio (a_{wc})
2. Determine the limiting slenderness ratio (λ_{rw})
3. Compare the web slenderness ratio to the limiting slenderness ratio
4. Determine R_b based on the comparison.

First, we need to calculate the web compactness ratio as below:

$$a_{wc} = \frac{2D_c t_w}{b_{f_c} t_{fc}} = \frac{2\times 115\times 7.5}{240\times 12} = 0.6 \tag{4.8}$$

where

- D_c = Depth of the web in compression
- t_w = Web thickness
- b_{fc} = Width of compression flange
- t_{fc} = Thickness of compression flange

then the limiting slenderness ratio λ_{rw} is calculated as below:

$$4.6\sqrt{\frac{E}{F_{yc}}} \le \lambda_{rw} = \left(3.1+\frac{5}{a_{wc}}\right)\sqrt{\frac{E}{F_{yc}}} \le 5.7\sqrt{\frac{E}{F_{yc}}} \tag{4.9}$$

where E is modulus of elasticity of steel, and F_{yc} is yield strength of the compression flange. By replacing values in the equations, we have:

$$4.6\sqrt{\frac{200000}{360}} \le \lambda_{rw} = \left(3.1+\frac{5}{0.6}\right)\sqrt{\frac{200000}{360}} \le 5.7\sqrt{\frac{200000}{360}}$$

$$108.5 \le \lambda_{rw} = 211.5 \le 134.4$$

Since the calculated λ_{rw} exceeds the upper limit, we use the upper limit, and λ_{rw} is 134.4. Once these two are determined, then the following Comparison is made:

- If $\frac{2D_c}{t_w} \le \lambda_{rw}$, then $R_b = 1.0$.
- If $\frac{2D_c}{t_w} > \lambda_{rw}$, calculate R_b using AASHTO equations (not provided in this example).

In case of Neka bridge and for longitudinal beams, we have:

$$\frac{2\times D_c}{t_w}=\frac{2\times 115}{7.5}=30.7<\lambda_{\mathrm{rw}}=134\rightarrow R_{\mathrm{b}}=1$$

Therefore, the web is not slender, and no load-shedding occurs, indicating that the full section capacity can be utilized without reduction due to web slenderness.

Next, local buckling resistance is checked. This step evaluates the compression flange's susceptibility to local buckling. The following steps must be taken:

1. Calculate the flange slenderness ratio (λ_{f})
2. Determine the compact (λ_{pf}) and noncompact (λ_{rf}) slenderness limits
3. Classify the flange based on its slenderness
4. Calculate the nominal flexural resistance (F_{nc}) based on flange classification

First, flange slenderness ratio is determined as shown below:

$$\lambda_{\mathrm{f}}=\frac{b_{\mathrm{fc}}}{2t_{\mathrm{fc}}}=\frac{240}{2\times 12}=10 \tag{4.10}$$

where all parameters are as defined before. Next, compact and noncompact slenderness limits must be determined as follows, with parameters defined as above:

$$\lambda_{\mathrm{Pf}}=0.38\sqrt{\frac{E}{F_{y_c}}}=0.38\sqrt{\frac{200000}{360}}=9 \tag{4.11}$$

$$\lambda_{\mathrm{rf}}=0.56\sqrt{\frac{E}{F_{y_c}}}=0.56\sqrt{\frac{200000}{360}}=13.2 \tag{4.12}$$

Now, it is possible to classify the flange based on its slenderness according to the below criteria:

- If $\lambda_{\mathrm{f}}\leq\lambda_{\mathrm{pf}}$: Compact flange
- If $\lambda_{\mathrm{pf}}\leq\lambda_{\mathrm{f}}\leq\lambda_{\mathrm{rf}}$: Noncompact flange
- If $\lambda_{\mathrm{f}}>\lambda_{\mathrm{rf}}$: Slender flange

In case of the Neka longitudinal beams, flange is classified as noncompact, therefore

$$\begin{aligned}F_{nc}&=F_{yc}\left[1-\left(1-\frac{F_{yr}}{F_{yc}}\right)\left(\frac{\lambda_f-\lambda_{pf}}{\lambda_{rf}-\lambda_{pf}}\right)\right]\\&=360\times\left[1-\left(1-\frac{0.7\times 360}{360}\right)\left(\frac{10-9}{13.2-9}\right)\right]=334.3\text{ MPa}\end{aligned} \tag{4.13}$$

where F_{yr} is compression-flange stress at the onset of nominal yielding, which is taken as the smaller of:

- $0.7\ F_{yc}$
- F_{yw} (yield strength of the web)

However, F_{yr} should not be smaller than $0.5\ F_{yc}$.

The local buckling resistance of the compression flange is 334.3 MPa. This value represents the maximum stress the flange can sustain before local buckling occurs, considering its noncompact classification.

Next, lateral–torsional buckling resistance is checked, which assesses the girder's resistance to lateral–torsional buckling, a global instability phenomenon, to ensure the girder maintains stability under lateral and torsional deformations, especially in regions of negative moment. The following steps must be taken:

1. Calculate limiting unbraced lengths L_p and L_r
2. Determine the nominal flexural resistance (F_{nc}) based on the actual unbraced length L_b
3. Consider moment gradient factor C_b if applicable
4. Apply any necessary modifies (e.g. R_b for web load-shedding)

First limiting unbraced lengths are determined as follows:

$$L_p = 1 \times r_t \times \sqrt{\frac{E}{F_{yc}}} \tag{4.14}$$

$$L_r = \pi \times r_t \times \sqrt{\frac{E}{F_{yr}}} \tag{4.15}$$

where r_t is the radius of gyration of the compression flange about the weak axis, and the rest of the parameters are as defined before. r_t is determined using the following equation:

$$r_t = \frac{b_{fc}}{\sqrt{12\left(1+\frac{D_c t_w}{3b_{fc}t_{fc}}\right)}} = \frac{240}{\sqrt{12\left(1+\frac{115\times 7.5}{3\times 240\times 11}\right)}} = 66\,\text{mm} \tag{4.16}$$

where D_c is the depth of the web in compression, and rest of the parameters are as defined before. Therefore, L_p and L_r are:

$$L_p = 1 \times r_t \times \sqrt{\frac{E}{F_{y_c}}} = 1 \times 66 \times \sqrt{\frac{200 \times 10^3}{360}} = 1556\,\text{mm}$$

$$L_r = \pi \times r_t \times \sqrt{\frac{E}{F_{y_r}}} = \pi \times 66 \times \sqrt{\frac{200 \times 10^3}{0.7 \times 360}} = 5838\,\text{mm}$$

F_{nc} depends on L_p and L_r, and is calculated as shown below:

- For $L_b \leq L_p : F_{nc} = R_b F_{yc}$
- For $L_p < L_b \leq L_r : F_{nc} = C_b \left[1 - \left(1 - \frac{F_{yr}}{R_h F_{yc}}\right)\left(\frac{L_b - L_p}{L_r - L_p}\right)\right] R_b R_h F_{yc}$
- For $L_b > L_r : F_{nc} = F_{cr} \leq R_b \mathrm{R}_h F_{yc}$

where

- C_b = Moment gradient factor (1.0 for uniform moment)
- R_b = Web load-shedding factor
- R_h = Hybrid factor (1.0 for homogeneous sections)

Given L_b is 3 m, F_{nc} is calculated as below:

$$F_{nc} = 1 \times \left[1 - \left(1 - \frac{0.7 \times 360}{360}\right)\left(\frac{3000 - 1556}{5838 - 1556}\right)\right] \times 1 \times 1 \times 360 = 323.7\ \text{MPa}$$

The lateral–torsional buckling resistance of the beam is 323.7 MPa. This value represents the maximum stress the beam can sustain before lateral–torsional buckling occurs, considering its unbraced length falls in the inelastic buckling range.

Next, stress control is conducted, which compares the factored moment (M_u) to the factored flexural resistance ($\phi_f M_n$). The steps to be taken are as follows:

1. Calculate the actual stress in the beam.
2. Determine the allowable stress (nominal flexural resistance).
3. Compare the actual stress to the allowable stress.

Actual stress in the beam is calculated as shown below:

$$f_a = \frac{M_u}{S} \tag{4.17}$$

$$M_{\mathrm{d(mid)}} = \frac{\omega L^2}{8} = \frac{14}{2} \times \frac{3^2}{8} = 7.9\,\mathrm{kN.m}$$

$$M_{\mathrm{u(mid)}} = 1.25\mathrm{M}_d + 1.75M_{L+1} = 1.25 \times 7.9 + 1.75 \times 138 = 251.4\,\mathrm{kN.m}$$

$$f_{\mathrm{a}} = \frac{251.4}{675000 \times 10^6} = 372\ \mathrm{MPa}$$

where M_{u} is the factored moment from applied loads, and S is the section modulus. Allowable stresses can be determined as minimum of nominal flexural resistance based on local buckling (F_{nc}^{LB}) and lateral–torsional buckling (F_{nc}^{LTB}), which were determined earlier. Therefore:

$$F_n = \min\left(F_{nc}^{LB}, F_{nc}^{LTB}\right) = \min(334.3, 323.7) = 323.7\ \mathrm{MPa} \tag{4.18}$$

Now we can compare actual and allowable stresses:

$$f_a = 372\,\mathrm{MPa} > F_n = 323.7\,\mathrm{MPa}$$

The stress control check reveals that the actual stress in the beam (372 MPa) exceeds the allowable stress (323.7 MPa). This indicates that the beam is overstressed according to the AASHTO LRFD 6.10.8 criteria. RF for inventory level in ULS is:

$$\mathrm{RF}_{\mathrm{inventory}} = \frac{323.7 \times 675000 \times 10^{-6} - 1.25 \times 7.9}{1.75 \times 138} = .86 < 1$$

This means it is not adequate in inventory level. Repeating it for operating limit state of ULS, we get:

$$\mathrm{RF}_{\mathrm{operating}} = \frac{218.5 - 1.25 \times 7.9}{1.35 \times 138} = 1.1 > 1$$

This means the longitudinal beam has enough capacity for operating level of ULS.

4.3.1.1.4 Shear Capacity

AASHTO LRFD 6.10.9 outlines the procedure for checking the shear capacity of steel I-girders. The process involves:

- calculating the factored shear demand,
- determining the shear buckling coefficient,
- checking the web slenderness ratio,
- calculating the shear buckling resistance, and finally
- comparing the shear demand to the shear resistance.

The procedure also includes calculating a rating factor to assess the adequacy of the girder for shear.

First, shear demand is calculated as shown as follows:

$$\Phi_2 = \frac{1.44}{\sqrt{L_\varphi} - 0.2} + 0.82 = \frac{1.44}{\sqrt{3 \times 2} - 0.2} + 0.82 = 1.46$$

$$V_{L+I\,(\text{end})} = 0.5 \times 0.8 \times 371.5 \times 1.46 = 217\,\text{kN}$$

$$V_{\text{D}\,(\text{end})} = \frac{14}{2} \times \frac{3}{2} = 10.5\,\text{kN}$$

$$V_{\text{U}(\text{end})} = 1.25 \times 10.5 + 1.75 \times 217 = 392.5\,\text{kN}$$

The shear buckling coefficient k is a critical parameter in assessing the shear capacity of structural elements, particularly in web panels of beams. This coefficient helps determine the susceptibility of a web to shear buckling under load. A higher k value indicates a greater resistance to shear buckling, which is essential for ensuring the stability and safety of structural components under various loading conditions. Shear buckling coefficient k is calculated using the below formula:

$$k = 5 + \frac{5}{\left(\frac{d_0}{D}\right)^2} = 5 + \frac{5}{\left(\frac{3000}{206}\right)^2} = 5 \tag{4.19}$$

where d_0 is the stiffener spacing (distance between vertical stiffeners), and D is depth of the web. Next, web slenderness ratio is determined, as shown below:

$$\frac{D}{t_w} = \frac{206}{7.5} = 27.5$$

Now, correction factor C can be determined based on the below logic:

$$C = \begin{cases} 1 & \text{if } \frac{D}{t_w} \le 1.12\sqrt{\frac{E_k}{F_{yw}}} \\ \frac{1.12}{\frac{D}{t_w}}\sqrt{\frac{E_k}{F_{yw}}} & 1.12\sqrt{\frac{E_k}{F_{yw}}} < \frac{D}{t_w} \le 1.40\sqrt{\frac{E_k}{F_{yw}}} \\ else & \text{if } \frac{D}{t_w} \ge 1.40\sqrt{\frac{E_k}{F_{yw}}} \end{cases} \tag{4.20}$$

where F_{yw} is yield strength of web steel. In our example, therefore, C is:

$$1.12\sqrt{\frac{Ek}{F_{yw}}} = 1.12\sqrt{\frac{200000\times5}{360}} = 59 > \frac{D}{t_w} = 27.5 \Rightarrow C = 1$$

And nominal shear resistance can be determined as follows, where all parameters are as defined above:

$$V_n = 0.58F_{yw}Dt_wC \tag{4.21}$$

$$V_n = 0.58\times360\times206\times7.5\times1\times10^{-3} = 322.6\,kN$$

Given this is less than shear demand of 392.2 kN, the longitudinal doesn't have enough shear capacity. RF can be calculated as shown below for inventory and operating levels of ultimate limit states. It is evident that bridge has an RF of less than 1 for inventory level, yet it is sufficient for operating level in ULS.

$$RF_{\text{inventory}} = \frac{322.6 - 1.25\times10.5}{1.75\times217} = 0.81 < 1$$

$$RF_{\text{operating}} = \frac{322.6 - 1.25\times10.5}{1.35\times217} = 1.06 > 1$$

4.3.1.2 Flexural Assessment: Cross Beams

Similar to longitudinal beams, the flexural assessment has to be repeated for cross beams. Neka bridge has 13 sets of cross beams of IPBL400, each with a length of 5.5 m. The critical positioning of design live load on longitudinal beams is presented in Figure 4.5. On each cross beam, there are two longitudinal beams resting 1.5 m apart, transferring loads to cross beams (Figures 4.6 and 4.7). Calculations are presented below with more brevity, as detailed calculations were presented for longitudinal beams.

4.3.1.2.1 Fatigue Resistance

Fatigue resistance for cross beams is calculated using the same formulas stated for longitudinal beam, as shown below:

$$R_{\text{LL(rail)}} = (1 + 0.47 + 0.47) \times 250 + 0.06 \times 80 = 489.8 \text{ kN.m}$$

$$R_{\text{LL+IM(rail)}} = 0.5 \times 0.8 \times 489.8 = 196 \text{ kN.m}$$

$$M_{\text{LL(mid)}} = 196 \times (1 + 1) = 392 \text{ kN.m}$$

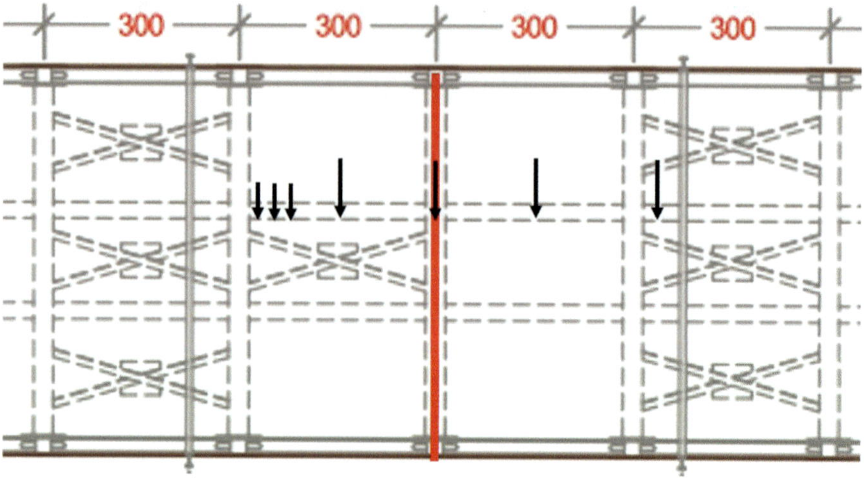

FIGURE 4.5 Cross beam under maximum code load.

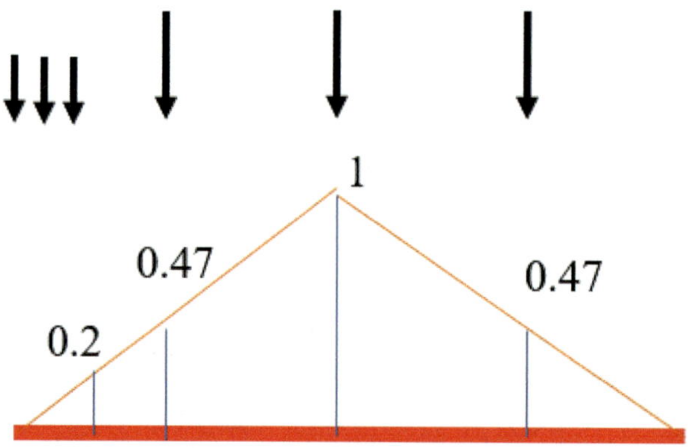

FIGURE 4.6 Influence line of support force (two longitudinal beams) under maximum code load.

$$\Phi_3 = \frac{2.16}{\sqrt{5.5 \times 2} - 0.2} + 0.73 = 1.42$$

$$M_{\mathrm{LL+IM(end)}} = 392 \times 1.42 = 556 \text{ kN.m}$$

$$N = 365 \times 75 \times 2 \times (40 \times 10) = 21{,}900{,}000$$

$$(\Delta F)_n = \left(\frac{82 \times 10^{11}}{21{,}900{,}000} \right)^{\frac{1}{3}} = 72.1\,\text{MPa}$$

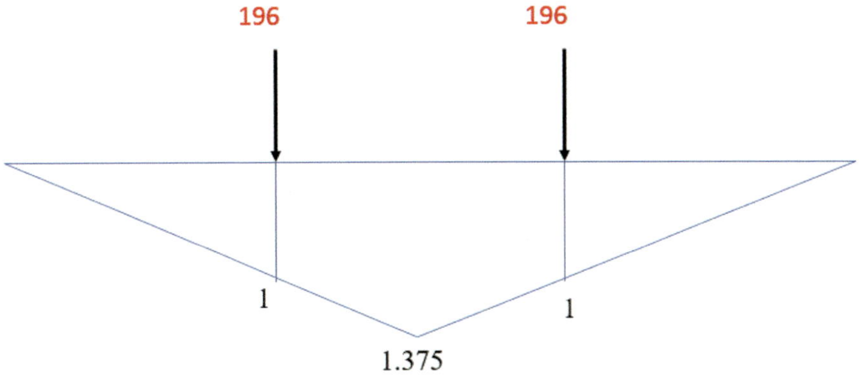

FIGURE 4.7 Influence line of bending moment at mid-span of the cross beam under maximum code load.

For Fatigue II limit state (limited cycles),

$$0.8 \times M_{\text{LL+IM(end)}} = 0.8 \times 556 = 445 \text{ kN.m}$$

$$\frac{445 \times 10^6}{2310000} = 192.6 > 72.1$$

$$\text{RF} = \frac{72.1}{192.6} = 0.37 < 1$$

For Fatigue I limit state (unlimited cycles),

$$1.75 \times M_{\text{LL+IM(mid)}} = 1.75 \times 556 = 973 \text{ kN.m}$$

$$\frac{973 \times 10^6}{2310000} = 421 > (\Delta F)_{\text{TH}} = 168$$

$$\text{RF} = \frac{168}{421} = 0.4 < 1$$

Therefore, cross beams are not adequate for fatigue under design load.

4.3.1.2.2 Flexural Capacity

Flexural capacity for cross beams is performed using the same formulas stated for longitudinal beam, as shown below:

$$M_{\text{D(end)}} = 42/2 \times (1+1) = 42 \text{ kN.m}$$

$$M_{\text{U(end)}} = 1.25 \times 42 + 1.75 \times 556 = 1025.5 \text{ kN.m}$$

$$\lambda_{rw} = 5.7\sqrt{\frac{E}{F_{y_c}}} \Rightarrow \lambda_{rw} = 5.7\sqrt{\frac{200000}{360}} = 134$$

$$\frac{2\times D_c}{t_w} = \frac{2\times 195}{11} = 35.5 < \lambda_{rw} = 134 \rightarrow R_b = 1$$

4.3.1.2.3 Local Buckling Resistance

$$\lambda_f = \frac{b_{f_c}}{2t_{fc}} = \frac{300}{2\times 19} = 7.9 < \lambda_{Pf} = 0.38\sqrt{\frac{E}{F_{yc}}} = 0.38\sqrt{\frac{200\times 10^3}{360}} = 9$$

$$\lambda_{pf} \geq \lambda_f \rightarrow F_{nc} = R_b R_h F_{y_c} = 1\times 360 = 360\,MPa$$

4.3.1.2.4 Lateral–Torsional Buckling Resistance

$$r_t = \frac{b_{fc}}{\sqrt{12\left(1+\frac{D_c t_w}{3b_{fc}t_{fc}}\right)}} = \frac{300}{\sqrt{12\left(1+\frac{195\times 11}{3\times 300\times 19}\right)}} = 81.6$$

$$L_P = 1\times r_t \times \sqrt{\frac{E}{F_{y_c}}} = 1\times 81.6\times\sqrt{\frac{200\times 10^3}{360}} = 1\times 81.6\times 23.57 = 1923 \text{ mm}$$

$$L_r = \pi\times r_t \times \sqrt{\frac{E}{F_{y_r}}} = \pi\times 81.6\times\sqrt{\frac{200\times 10^3}{0.7\times 360}} = \pi\times 81.6\times 28.17 = 7218\,\text{mm}$$

$$L_p < L_b = 2000 < L_r \rightarrow$$

$$F_{nc} = 1\times\left[1-\left(1-\frac{0.7\times 360}{360}\right)\left(\frac{2000-1923}{7218-1923}\right)\right]\times 1\times 360 = 358 \leq 1\times 360$$

4.3.1.2.5 Stress Control

$$\frac{1025.5\times 10^6}{2310000} = 443.9 > \min(360, 358) = 358\,\text{MPa}$$

$$\mathrm{RF}_{\text{inventory}} = \frac{358 \times 2310000 \times 10^{-6} - 1.25 \times 42}{1.75 \times 556} = 0.8 < 1$$

$$\mathrm{RF}_{\text{operating}} = \frac{831.6 - 1.25 \times 42}{1.35 \times 556} = 1.03 > 1$$

Therefore, flexural capacity is adequate for the operating level under the design load for cross beams.

4.3.1.2.6 Shear Capacity

Shear capacity for cross beams is performed using the same formulas stated for the longitudinal beam, as shown below:

$$V_{\text{L+I (end)}} = 196 \times 1.28 = 250.9 \text{ kN}$$

$$\Phi_2 = \frac{1.44}{\sqrt{5.5 \times 2} - 0.2} + 0.82 = 1.28$$

$$V_{\text{D(end)}} = 42/2 = 21 \text{ kN}$$

$$V_{\text{U(end)}} = 1.25 \times 21 + 1.75 \times 250.9 = 465.3 \text{ kN}$$

$$k = 5 + \frac{5}{\left(\frac{d_0}{D}\right)^2} = 5 + \frac{5}{\left(\frac{5500}{390}\right)^2} = 5$$

$$\frac{D}{t_w} = \frac{390}{11} = 35.5$$

$$1.12\sqrt{\frac{Ek}{F_{yw}}} = 1.12\sqrt{\frac{200000 \times 5}{360}} = 59 > \frac{D}{t_w} \Rightarrow C = 1$$

$$V_n = 0.58 \times 360 \times 390 \times 11 \times 1 \times 10^{-3} = 896 \text{ kN} > 465.3$$

$$\mathrm{RF}_{\text{inventory}} = \frac{896 - 1.25 \times 21}{1.75 \times 250.9} = 1.98 > 1$$

Therefore, shear capacity is adequate for cross members in the ULS inventory level under design load.

4.3.1.3 Axial Assessment: Top Chord Axial Members

The critical member of the upper chord is located at the mid-span with the highest compressive stress (shown in red in Figure 4.8). By using the influence line of the truss or through truss analysis via the section or joint method, the maximum forces of the live and dead loads on the horizontal member of the upper chord truss at mid-span can be calculated as shown below:

$$P_{\text{LL(up-mid)}} = 0.8\times 0.5\times\begin{bmatrix}240\times 2\times(-0.44-1.01-1.53)+239.5\times 2\times(-2.06)\\ +371.2\times 2\times(-2.57)+466.7\times(2.43)\end{bmatrix}$$
$$= -2184\,\text{kN}$$

$$P_{\text{D(up-mid)}} = (14\times 3)\times 0.5$$
$$\times\left[2\times(-0.44-1.01-1.53)+2\times(-2.06)+2\times(-2.57)+(-2.43)\right]$$
$$= -371\text{kN}$$

$$\Phi_3 = \frac{2.16}{\sqrt{36}-0.2}+0.73 = 1.10$$

$$P_{\text{U(up-mid)}} = 1.25\times(-371)+1.75\times(-2184)\times 1.1 = -4668\,\text{kN}$$

4.3.1.3.1 Fatigue Resistance

Fatigue detail category C is considered for this element. Also, it is assumed that every day 40 trains cross over the bridge, each creating a loading cycle. Fatigue resistance can be calculated using the same formulas shown before:

$$N = 365\times 75\times 1\times(40) = 1{,}095{,}000$$

$$(\Delta F)_n = \left(\frac{14.4\times 10^{11}}{1{,}095{,}000}\right)^{\frac{1}{3}} = 109.6\text{ MPa}$$

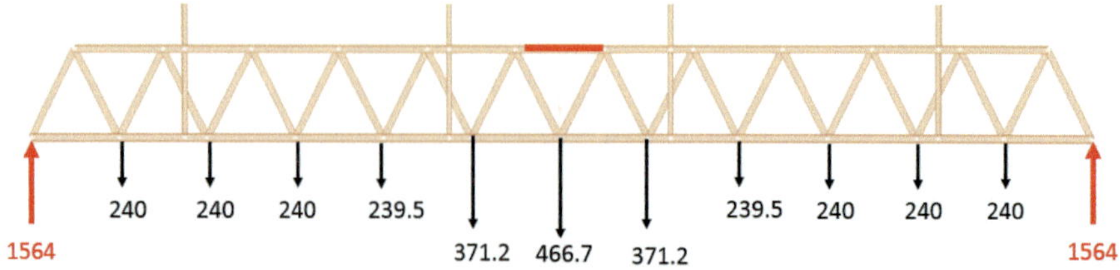

FIGURE 4.8 Neka truss bridge under standard design load, and upper chord in mid-span with maximum compressive stresses.

For fatigue II limit state (limited cycle)

$$0.8 \times P_{LL+IM(mid)} = 0.8 \times 2402 = 1921.6\,\text{kN}$$

$$\frac{1921.6 \times 10^3}{14030} = 137 > 109.6$$

$$\text{RF} = \frac{109.6}{137} = 0.8 < 1$$

Therefore, horizontal compressive members are inadequate for fatigue under design loads.

4.3.1.3.2 Axial Compressive Capacity

In accordance with AASHTO LRFD 6.9.4 [1], the axial load capacity of structural members is determined by calculating the radius of gyration, the critical buckling load, and the nominal axial load capacity. The process involves comparing the actual load to the calculated resistance to ensure that the structure can safely support the applied loads. The steps to check the axial load capacity are as follows:

1. Calculate the radius of gyration in both *x* and *y* directions (r_x and r_y).
2. Calculate the nominal axial load capacity (P_0).
3. Calculate the Euler's critical buckling load (P_e).
4. Check the ratio of P_0 to P_e.
5. Calculate the nominal axial load capacity (P_n).
6. Determine the reduced axial load capacity (P_r).
7. Calculate the rating factor (RF).

The radius of gyration can be calculated using the following formulas (also given in the beginning of Section 4.3):

$$r_x = \sqrt{\frac{I_x}{A}} = \sqrt{\frac{9656}{140.3}} = 83\,\text{mm}$$

$$r_y = \sqrt{\frac{I_y}{A}} = \sqrt{\frac{9637}{140.3}} = 83\,\text{mm}$$

where I_x and I_y are moment of inertia about the *x*- and *y*-axis, respectively, and *A* is the cross-sectional area of the member. Next, nominal axial load capacity (P_0) can be calculated using the below formula:

$$P_0 = 1 \times F_y \times A = 1 \times 360 \times 14030 \times 10^{-3} = 5051\,\text{kN} \quad (4.22)$$

Euler's critical buckling load (P_e) is calculated next:

$$P_e = \frac{\pi^2 E}{\left(\frac{L_{\text{eff}}}{r}\right)^2} \times A = \frac{\pi^2 \times 200000}{\left(\frac{7530}{83}\right)^2} \times 14030 \times 10^{-3} = 3361\,\text{kN} \quad (4.23)$$

Therefore:

$$n = \frac{P_0}{P_e} = \frac{5051}{3361} \approx 1.5 < 2$$

This indicates that the column is not slender. Therefore, the nominal axial capacity (P_n) and reduced axial capacity (P_r) are:

$$P_n = 0.658^n \times P_0 = 0.658^{1.5} \times 5051 = 2696\,\text{kN} \quad (4.24)$$

$$P_r = 0.9 \times P_n = 0.9 \times 2696 = 2426\,\text{kN} \quad (4.25)$$

Finally, RF for inventory and operating levels are:

$$\text{RF}_{\text{inventory}} = \frac{P_r - 1.25 \times M_D}{1.75 \times M_D} = \frac{2426 - 1.25 \times 371}{1.75 \times 2402} = 0.47 < 1$$

$$\text{RF}_{\text{operating}} = \frac{P_r - 1.25 \times M_D}{1.35 \times M_D} = \frac{2426 - 1.25 \times 371}{1.35 \times 2402} = 0.61 < 1$$

Therefore, the axial compressive members are not adequate at neither inventory nor operating levels under design load.

4.3.1.4 Axial Assessment: Horizontal Tensile Members

The critical member of the lower chord is located at the mid-span with the highest tensile stress (shown in red in Figure 4.9). By using the influence line of the truss or through truss analysis via the section or joint method, the maximum forces of the live and dead loads on the horizontal member of the lower chord truss at mid-span can be calculated as shown below.

$$\begin{aligned} P_{\text{LL(low-mid)}} &= 0.8 \times 0.5 \times \begin{bmatrix} 240 \times (0.57 + 0.49 + 1.08 + 0.91 + 1.66 + 1.4) \\ +239.5 \times (2.18 + 1.85) + 371.2 \times (2.78 + 2.35) + 466.7 \times (2.22) \end{bmatrix} \\ &= 2149\,\text{kN} \end{aligned}$$

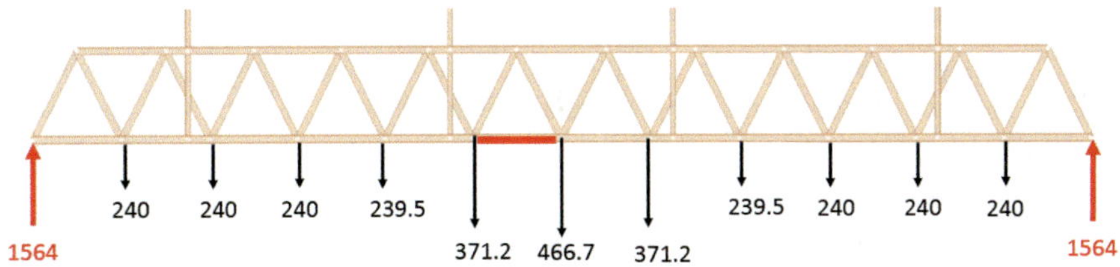

FIGURE 4.9 Neka truss bridge under standard design load, and lower chord in mid-span with maximum tensile stresses.

$$P_{\text{D(low-mid)}} = (14 \times 3) \times 0.5 \times \begin{bmatrix} (0.57 + 0.49 + 1.08 + 0.91 + 1.66 + 1.4) \\ +(2.18 + 1.85) + (2.78 + 2.35) + (2.22) \end{bmatrix}$$
$$= 367\,\text{kN}$$

$$\Phi_3 = \frac{2.16}{\sqrt{36} - 0.2} + 0.73 = 1.10$$

$$P_{\text{U(low-mid)}} = 1.25 \times (367) + 1.75 \times (2149) \times 1.1 = 4596\,\text{kN}$$

4.3.1.4.1 Fatigue Resistance

Fatigue resistance is calculated using the same methodology discussed before, as shown below:

$$N = 365 \times 75 \times 1 \times (40) = 1{,}095{,}000$$

$$(\Delta F)_n = \left(\frac{14.4 \times 10^{11}}{1{,}095{,}000} \right)^{\frac{1}{3}} = 109.6\,\text{MPa}$$

For fatigue II limit state (limited cycles)

$$0.8 \times P_{\text{LL+IM(mid)}} = 0.8 \times 2363.9 = 1891\,\text{kN}$$

$$\frac{1891 \times 10^3}{14030} = 135 > 109.6$$

$$\text{RF} = \frac{109.6}{135} = 0.81 < 1$$

Therefore, tensile axial members are not adequate in fatigue loading under design load.

4.3.1.4.2 Axial Tensile Capacity

Axial capacity a member can be calculated according to section 6.8.2 of AASHTO [3]. The factored tensile resistance, P_r, shall be taken as the minimum of the values from the below two equations:

$$P_r = \phi_y \times P_{ny} = \phi_y F_y A_g \tag{4.26}$$

$$P_r = \phi_u \times P_{nu} = \phi_u F_u A_n R_p U \tag{4.27}$$

in which

- P_{ny} is the nominal tensile resistance for yielding in gross section,
- F_y the specified minimum yield strength,
- A_g the gross cross-sectional area of the member,
- F_u the tensile strength,
- A_n are of the member as specified in clause 6.8.3, [1]
- R_p the reduction factor for holes taken equal to 0.90 for both holes punched full size and 1.0 for bolt holes drilled full size or sub-punched and reamed to size,
- U the reduction factor to account for shear lag, 1.0 for components in which force effects are transmitted to all elements, and as specified in article 6.8.2.2 for other cases, [1]
- ϕ_y the resistance factor for yielding of tension members as specified in article 6.5.4.2 [1]
- ϕ_u the resistance factor for fracture of tension members as specified in article 6.5.4.2 [1].

Therefore,

$$P_r = 0.95 \times 360 \times 14030 \times 10^{-3} = 4536\,\text{kN}$$

$$P_r = 0.8 \times 540 \times 8800 \times 0.9 \times 0.9 = 3079\ \text{kN}$$

$$P_r = 3079\ \text{kN} < 4596$$

$$\text{RF}_{\text{inventory}} = \frac{3079 - 1.25 \times 367}{1.75 \times 2364} = 0.63 < 1$$

$$\text{RF}_{\text{operating}} = \frac{3079 - 1.25 \times 367}{1.35 \times 2364} = 0.82 < 1$$

As evident, tensile axial members fail both at the inventory and operating levels under design load.

4.3.1.5 Axial Assessment: Diagonal Compressive Members

The critical diagonal member with compressive axial loads is located next to the support, as shown in Figure 4.10, which has a length of 3.3 m. By using the influence line of the truss or through truss analysis via the section or joint method presented in Figure 4.11, the maximum forces of the live and dead loads in this member can be calculated as shown below:

$$0.891 \times F_{\text{diagonal}} = 1564 \rightarrow F_{\text{diagonal}} = 1755.3\ \text{kN}$$

$$P_{\text{LL+IM}} = 0.8 \times 0.5 \times -1755.3 \times 1.1 = -772\,\text{kN}$$

$$P_{\text{DL}} = -126 / 0.891 = -141\,\text{kN}$$

$$\Phi_3 = \frac{2.16}{\sqrt{36} - 0.2} + 0.73 = 1.10$$

$$P_{\text{U}} = 1.25 \times (-141) + 1.75 \times (-772) = -1527\,\text{kN}$$

4.3.1.5.1 Fatigue Resistance

Fatigue resistance for compressive diagonal members is assessed in detail category A, as shown below:

$$N = 365 \times 75 \times 1 \times (40) = 1{,}095{,}000$$

$$(\Delta F)_n = \left(\frac{82 \times 10^{11}}{21{,}900{,}000}\right)^{\frac{1}{3}} = 72.1\,\text{MPa}$$

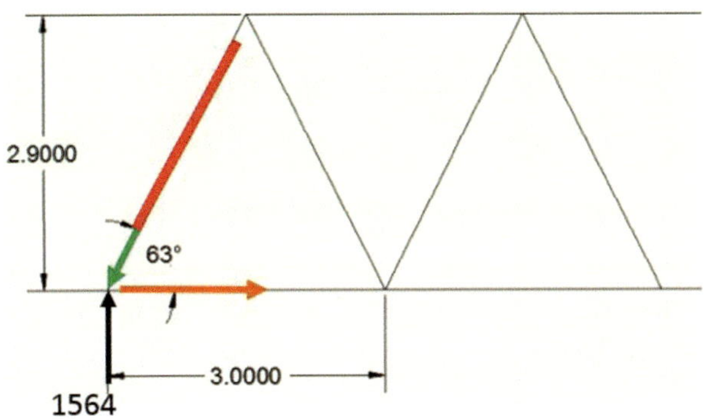

FIGURE 4.10 Neka truss bridge under standard design load, and compressive diagonal member at the support.

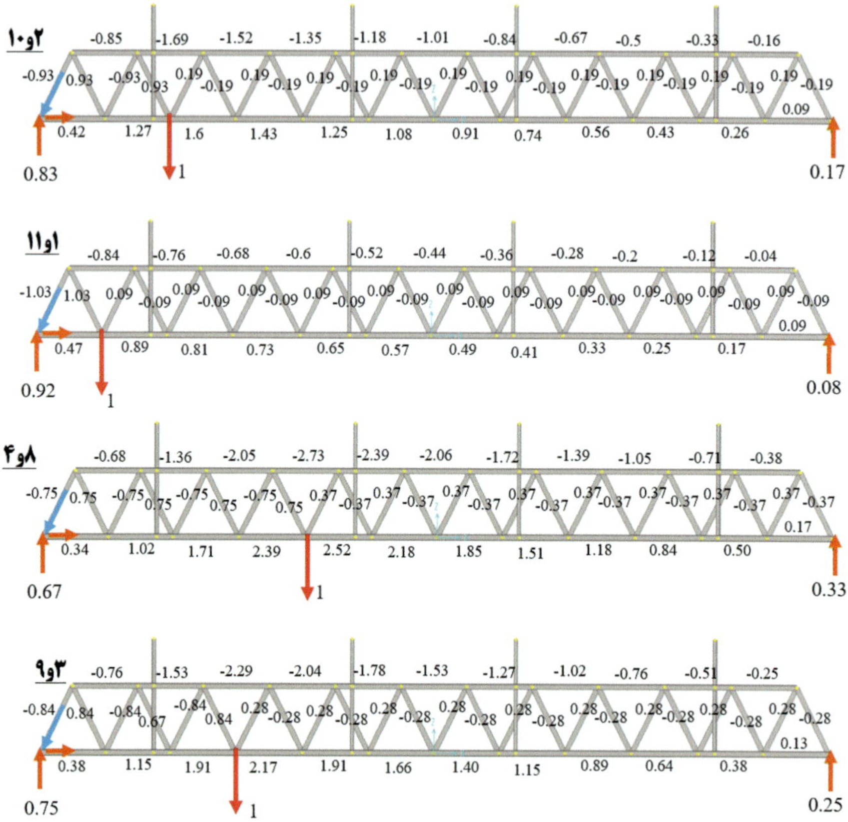

FIGURE 4.11 Influence line of the truss for a unit load.

For fatigue II limit state (limited cycles),

$$0.8 \times P_{\mathrm{LL+IM}} = 0.8 \times 772 = 617.6\,\mathrm{kN}$$

$$\frac{617.6 \times 10^3}{6430} = 96 > 72.1$$

$$\mathrm{RF} = \frac{72.1}{96} = 0.75 < 1$$

Therefore, the diagonal compressive axial members are not adequate for fatigue loading under design load.

4.3.1.5.2 Axial Compressive Capacity

Axial compressive capacity for diagonal members can be assessed using the same formulas discussed before:

$$r_x = \sqrt{\frac{I_x}{A}} = \sqrt{\frac{54100000}{6430}} = 91.7\,\text{mm}$$

$$r_y = \sqrt{\frac{I_y}{A}} = \sqrt{\frac{19500000}{6430}} = 55.1\,\text{mm}$$

$$P_0 = 1 \times F_y \times A = 1 \times 360 \times 6430 \times 10^{-3} = 2314.8\,\text{kN}$$

$$P_e = \frac{\pi^2 E}{\left(\frac{L_{\text{eff}}}{r}\right)^2} \times A = \frac{\pi^2 \times 200000}{\left(\frac{3300}{55.1}\right)^2} \times 6430 \times 10^{-3} = 3534.9\,\text{kN}$$

$$n = \frac{P_0}{P_e} = \frac{5051}{3361} \approx 0.65 < 2$$

Therefore, this member is not slender, hence:

$$P_n = 0.658^n \times P_o = 0.658^{0.65} \times 2314.8 = 1763\,kN$$

$$P_r = 0.9 \times 1763 = 1587kN > 1527$$

$$\text{RF}_{\text{inventory}} = \frac{P_r - 1.25 \times M_D}{1.75 \times M_D} = \frac{1587 - 1.25 \times 141}{1.75 \times 772} = 1.04 > 1$$

Therefore, this member has adequate compressive axial capacity for the inventory level under design load (hence operating level is not required to be assessed).

4.3.1.6 Axial Assessment: Diagonal Tensile Members

The critical diagonal member with tensile axial loads is located next to the support, as shown in Figure 4.12. By using the influence line of the truss or through truss analysis via the section or joint method, the maximum forces of the live and dead loads in this member can be calculated as shown below.

$$P_{\text{LL+IM}} = 0.8 \times 0.5 \times 1755.3 \times 1.1 = 772\,\text{kN}$$

$$P_{\text{DL}} = 126 / 0.891 = 141\,\text{kN}$$

$$\Phi_3 = \frac{2.16}{\sqrt{36 - 0.2}} + 0.73 = 1.10$$

$$P_{\text{U}} = 1.25 \times (141) + 1.75 \times (772) = 1527.3\,\text{kN}$$

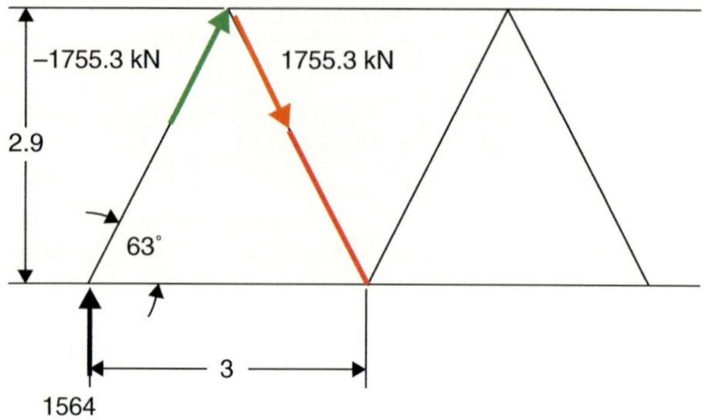

FIGURE 4.12 Neka truss bridge under standard design load, and tensile diagonal member at the support.

4.3.1.6.1 Fatigue Resistance

Fatigue resistance is assessed for detail category A, as shown below:

$$N = 365 \times 75 \times 1 \times (40) = 1{,}095{,}000$$

$$(\Delta F)_n = \left(\frac{82 \times 10^{11}}{21{,}900{,}000} \right)^{\frac{1}{3}} = 72.1MPa$$

$$0.8 \times P_{\text{LL+IM}} = 0.8 \times 772 = 617.6\,\text{kN}$$

$$\frac{617.6 \times 10^3}{6430} = 96 > 72.1$$

$$\text{RF} = \frac{72.1}{96} = 0.75 < 1$$

Therefore, the diagonal tensile axial members are not adequate for fatigue loading under design load.

4.3.1.6.2 Axial Tensile Capacity

Axial capacity of diagonal members under tension can be assessed using the same method discussed above, as shown below:

$$P_{\text{r}} = 1.25 \times (141) + 1.75 \times (772) = 1527.3\,\text{kN}$$

$$P_{\text{r}} = 0.95 \times 360 \times 6430 \times 10^{-3} = 4536\,\text{kN}$$

$$P_r = 0.8 \times 540 \times (6430 - 4 \times 22 \times 11) \times 0.9 \times 0.9 = 1911\,\text{kN}$$

$$P_r = 1911\,\text{kN} > 1527.3$$

$$\text{RF}_{ULS} = \frac{1911 - 1.25 \times 141}{1.75 \times 772} = 1.28 > 1$$

Therefore, this member has adequate tensile axial capacity for the ultimate limit state under design load (hence SLS is not required).

4.3.1.7 Summary of Load Rating for Design Loads

After calculating the RF for Neka bridge, a summary of these results for both flexural and axial elements is presented in Tables 4.3 and 4.4. Table 4.3 highlights the load rating factors for flexural members, including longitudinal and cross beams. Fatigue appears to be a significant issue for both elements, with low RF values of 0.47 and 0.40 for the longitudinal and cross beams, respectively. This suggests that the bridge components experience significant stress over time, which could lead to fatigue-related issues. The rating factors for bending and shear are slightly higher at inventory level, but they still indicate areas of concern, particularly in bending for the cross beam (0.80). Operating rating factors, on the other hand, are generally acceptable, with bending and shear RF values above 1.0, showing that under operating loading conditions, these beams perform adequately.

Table 4.4 presents the load rating factors for axial members, including the top and bottom chords and diagonal elements in both compression and tension. Similar to the

TABLE 4.3
Summary of load rating factors for flexural members of Neka Bridge

Element	Fatigue	Bending inventory	Bending operating	Shear inventory	Shear operating
Longitudinal beam	0.47	0.86	1.10	0.81	1.06
Cross beam	0.40	0.80	1.03	1.98	N.A.

TABLE 4.4
Summary of load rating factors for axial members of Neka Bridge

Element	Fatigue	Compressive/tensile (inventory)	Compressive/tensile (operating)
Top chord	0.80	0.47	0.61
Bottom chord	0.81	0.63	0.82
Diagonal comp.	0.75	1.04	N.A.
Diagonal tens.	0.75	1.28	N.A.

flexural members, fatigue remains a primary concern, with fatigue RF values below 1.0 for all members. The inventory RF values for the top chord indicate a substantial issue, particularly with a low compressive/tensile factor of 0.47. This suggests that, under extreme loads, the top chord may be highly susceptible to stress, which could impact overall stability. In contrast, the operating level factors are generally higher, particularly for the bottom chord (0.82) and diagonals in tension (1.28), implying that, under regular operational loads, these elements maintain sufficient capacity.

When a bridge assessment reveals low RF values, as seen with the Neka bridge, the next step is to refine the analysis to validate the initial results. Typically, bridge assessments are based on conservative assumptions, which can contribute to lower calculated RF values. However, the continued safe operation of the bridge often suggests that it possesses hidden capacity not accounted for in conventional assessments. Therefore, it is essential to verify these results before proceeding with strengthening or remediation.

As shown in Figure 4.1, the recommended next step is to incorporate the results of a load test if such data is available. Load testing provides real-world data on the bridge's response to controlled loads, allowing engineers to assess whether the conservative assumptions made in the original assessment were justified or overly cautious.

Since the Neka bridge was load-tested, with the results documented at the end of Chapter 3, this data will be used to refine the load rating assessment further. This approach helps ensure that any necessary interventions are based on accurate, field-verified data, potentially reducing the need for costly or extensive remediation efforts.

4.3.2 Load Rating with Field Load Testing

Field load testing helps reveal additional, often unaccounted for, structural capacity that can be leveraged, especially important for older bridges or those with unique construction features. In practice, load testing measures how a bridge reacts to controlled loads. It's typically kept within the elastic behaviour range, the load level where the structure can recover its original shape after the load is removed. This ensures safety during testing and provides realistic data under actual operating conditions.

The main goal is to compare measured responses with analytically predicted responses to decide if the bridge can safely carry more load than originally calculated. The following formulas define how the test results can adjust the bridge's load rating.

The load rating factor with testing adjustment (RF_T), noted here as RF_T, is calculated by adjusting the initial analytical rating (RF_C, calculated in previous section) based on load test data [3]:

$$RF_T = K \times RF_C \tag{4.28}$$

where:

- RF_T is the load rating factor after incorporating test results. A higher RF_T means a bridge can carry more load safely.
- RF_C is the load rating factor determined from analytical calculations, which is usually conservative.

- K is the adjustment coefficient derived from the test results. It accounts for differences between the bridge's real-life performance and its modelled behaviour.

In this way, the RF_T provides a more accurate picture of the bridge's load capacity by factoring in real-world data. If the test reveals that the bridge has extra capacity, the RF_T will be higher than the initial RF_C, reflecting this additional strength.

The adjustment coefficient, K, bridges the gap between measured data and analytical predictions. This coefficient acts as a multiplier, adjusting RF_C based on the load test outcomes:

$$K = 1 + k_a k_b \tag{4.29}$$

In this equation:

- k_a represents the section effect coefficient: a factor showing how the measured strain (deformation) compares to the calculated strain under the load.
- k_b , the extrapolation coefficient, considers whether test results can be extended beyond the test conditions to general operational scenarios.

Thus, K adjusts the analytical rating by combining two insights, the difference between calculated and actual strain, indicating any hidden load capacity, and how much we can extrapolate the test results, helping decide whether to trust the test results under higher loads. When $K > 1$, it indicates additional load-carrying potential, which increases the RF_T.

The section effect coefficient, k_a, quantifies the relationship between the bridge's theoretical and real-world strain under load. It measures how much of the bridge's material strength is actually being used compared to predictions:

$$k_a = \frac{\epsilon_C}{\epsilon_T} - 1 \tag{4.30}$$

where

- ϵ_T is the measured strain during the load test. Strain here represents how much the bridge material deforms under the load.
- ϵ_C is the calculated strain based on the original load rating analysis.

If ϵ_T (real strain) is less than ϵ_C (calculated strain), k_a becomes positive, signalling that the bridge materials are deforming less than expected under the load. This suggests extra strength that the model didn't account for. When k_a is positive, it shows the bridge may have hidden load-bearing capacity, allowing K to be greater than 1. Conversely, if ϵ_T is close to or exceeds ϵ_C , then k_a will be low or negative, which will keep K closer to 1, meaning the analytical rating was accurate or even unconservative.

The extrapolation coefficient, k_b, factors in whether the measured load test results can reliably predict behaviour under even greater loads. This is particularly important for translating test results into an operational context, as safety margins vary depending on how well we believe test results apply to real conditions.

The value of k_b is chosen based on the test load as a fraction of the evaluation load and whether the test result is extrapolatable to 1.33 times the evaluated load. If the test results can be extrapolated:

- If the test load / evaluation load ≤ 0.4: k_b =0 (no extrapolation assumed).
- If 0.4 < test load / evaluation load ≤ 0.7: k_b =0.8.
- If the test load / evaluation load > 0.7: k_b =1.

If the test results cannot be extrapolated:

- If test load / evaluation load ≤ 0.7: k_b =0 (results do not contribute to raising the rating).
- If test load / evaluation load > 0.7: k_b =0.5.

When test results are extrapolatable, we assume the bridge can handle a load up to 1.33 times what was tested. For example, if the test load represents 50 percent of the evaluation load, and the results extrapolate well, then we have confidence in raising the rating. If the test results cannot be extrapolated, we're more conservative. For instance, if the bridge was loaded to 70 percent or more of the evaluation load but cannot extrapolate to 1.33 times, k_b will be smaller, limiting K and reflecting caution in increasing the load rating. These are summarized in Table 4.5.

TABLE 4.5
k_b values from MBE [3]

k_b	Test Load / evaluation load (T / W)	Can test results be extrapolated to 1.33 × evaluation load?
0.0	$\frac{T}{W} \le 0.4$	Yes
0.8	$0.4 < \frac{T}{W} \le 0.7$	Yes
1.0	$\frac{T}{W} > 0.7$	Yes
0.0	$\frac{T}{W} \le 0.7$	No
0.5	$\frac{T}{W} > 0.7$	No

In the following sections, the modified rating factor, RF_T, will be calculated for bridge members that did not meet the initial rating factor requirements. Members that have already passed and demonstrated sufficient capacity will not require reassessment. However, for some cases, assessments will still be conducted on these members for purposes of comparison and to enhance the knowledge and understanding of readers.

4.3.2.1 Flexural Assessment: Longitudinal Beams

4.3.2.1.1 Fatigue Resistance

From Section 4.3.1.1, the maximum bending under design load is 138 kN.m (with dynamic load factor applied). Also, from 4.3.1.1.1 fatigue check, RF was calculated to be 0.47. During dynamic load testing, the maximum bending under train formation 1 (single locomotive) and 2 (single locomotive with two fully loaded wagons) were 69 and 71.5 kN.m, respectively. *T/W* is approximately 0.5, and hence k_b is 0.80. As shown in Figure 3.45, maximum strain in top chord in the middle of the span under train formation 1 was 220 μstrain, which equals 44 MPa.

Miner's rule, also known as the Palmgren–Miner rule, is a method used to calculate cumulative fatigue damage in materials subjected to varying stress levels. The law states that fatigue failure occurs when the sum of the ratios of the number of cycles at each stress level to the number of cycles to failure at that stress level reaches 1. The formula for Miner's rule is:

$$D = \Sigma\left(\frac{n_i}{N_i}\right) \tag{4.31}$$

where D is cumulative damage, n_i is the number of cycles at stress level i, and N_i is number of cycles to failure at stress level i. The R value, or remaining life ratio, is calculated as:

$$R = 1 - D \tag{4.32}$$

in which R represents the fraction of fatigue life remaining. An R value of 1 means no fatigue damage has occurred, while an R value of 0 indicates complete fatigue failure.

The number of fatigue cycles is determined using the rainflow counting method, based on strain measurements with ranges of 44, 12, and 6 MPa according to Figure 3.45. In this method, the rainflow counting technique helps to identify individual load cycles, and the damage from each cycle is summed using Miner's rule to estimate the remaining fatigue life as shown below:

$$\sigma_{\mathrm{T1}} = E \times \sigma_{T1} = 2\times 10^5 \times 220 \times 10^{(-6)} = 44\,\mathrm{MPa}$$

$$\sigma_{T2} = 0.2 \times 60 = 12 \;\; \mathrm{MPa}$$

$$\sigma_{T3} = 0.2 \times 30 = 6 \text{ MPa}$$

At time of the assessment, the bridge had been in service for 25 years, and it was assumed that each day 20 trains, each with 10 wagons (with two bogies), crossed over it. Also this member is considered detail category A. Therefore:

$$n = 365 \times 25 \times 2 \times (20 \times 10) = 3{,}650{,}000$$

$$(\Delta F)_n = \left(\frac{82 \times 10^{11}}{N}\right)^{\frac{1}{3}}$$

$$44 = \left(\frac{82 \times 10^{11}}{N}\right)^{\frac{1}{3}} \Rightarrow N_1 = 9.6 \times 10^7$$

$$12 = \left(\frac{82 \times 10^{11}}{N}\right)^{\frac{1}{3}} \Rightarrow N_2 = 4.7 \times 10^9$$

$$6 = \left(\frac{82 \times 10^{11}}{N}\right)^{\frac{1}{3}} \Rightarrow N_3 = 3.8 \times 10^{11}$$

$$R = 1 - D = 1 - \left(\frac{n_1}{N_1} + \frac{n_2}{N_2} + \frac{n_3}{N_3}\right) = 1 - \frac{3.65 \times 10^6}{9.6 \times 10^7} + \frac{3.65 \times 10^6}{4.7 \times 10^9} + \frac{3.65 \times 10^6}{3.8 \times 10^{11}} = 0.96$$

This high R value (0.96) suggests that the current loading conditions are causing minimal fatigue damage, and the structure is likely to perform well for a significant period.

4.3.2.1.2 Flexural Capacity

First the stress in this member is calculated as shown below:

$$\sigma_C = \frac{M}{S} = \frac{69 \times 10^6}{675000} = 102\, MPa$$

where M is the moment in N.mm, and S is section modulus of longitudinal beams in mm³. Therefore:

$$\varepsilon_C = \frac{\sigma_C}{E} = \frac{102}{200000} \times 10^6 = 510\, \mu\varepsilon$$

$$k_a = \frac{\epsilon_C}{\epsilon_T} - 1 = \frac{510}{220} - 1 = 1.3$$

$$K = 1 + k_a k_b = 1 + 1.3 \times 0.8 = 2$$

From 4.3.1.1 flexural capacity, we have $RF_{inventory}$ is 0.86, therefore:

$$RF_T = K \times RF_C = 2 \times 0.86 = 1.72$$

Hence, Neka's stringers have adequate flexural capacity in inventory level using bridge's load test results. Although not required, the same is repeated for operating level for readers benefit. From 4.3.1.1, we have $RF_{operating}$ is 1.1, therefore:

$$RF_T = K \times RF_C = 2 \times 1.1 = 2.2$$

This demonstrates the benefit of performing field load testing, as it further evidences the hidden capacity of the bridge in sustaining operational loads.

4.3.2.2 Flexural Assessment: Cross Beams

4.3.2.2.1 Fatigue Resistance

From Section 4.3.1.2 fatigue calculations, the maximum bending under design load is 566 kN.m (with dynamic load factor applied). Also, RF was calculated to be 0.40. During dynamic load testing, the maximum bending under train formation 1 (single locomotive) and 2 (single locomotive with two fully loaded wagons) were 333 kN.m and 369 kN.m, respectively. T/W is approximately 0.59 (or 0.65), and hence k_b is 0.80. As shown in Figure 3.48, maximum strain in top chord in the middle of the span under train formation 1 was 600 μ strain, which equals 120 MPa.

The remaining fatigue life is calculated based on Miner's rule. The member is considered detail category A. The number of fatigue cycles is determined using the rainflow counting method, based on strain measurements with a range of 600 μ strain:

$$\sigma_{T1} = \mathrm{E} \times \sigma_{T1} = 2 \times 10^5 \times 600 \times 10^{-6} = 120\,\mathrm{MPa}$$

$$n_1 = 365 \times 25 \times 2 \times (20 \times 10) = 3{,}650{,}000$$

$$(\Delta F)_n = \left(\frac{82 \times 10^{11}}{N}\right)^{\frac{1}{3}}$$

$$120 = \left(\frac{82 \times 10^{11}}{N}\right)^{\frac{1}{3}} \Rightarrow N_1 = 4.7 \times 10^6$$

$$R = 1 - D = 1 - \left(\frac{n_1}{N_1}\right) = 1 - \frac{3.65 \times 10^6}{4.7 \times 10^6} = 0.22$$

An R value of 0.22 is generally considered poor and indicates that the majority of the fatigue life has been consumed. This typically triggers a detailed evaluation of the component's condition and consideration of appropriate actions to manage the fatigue risk.

4.3.2.2.2 Flexural Capacity

Flexural capacity for cross beams is calculated using the formula and values explained above:

$$\sigma_C = \frac{M}{S} = \frac{333 \times 10^6}{231000} = 144\,\text{MPa}$$

$$\varepsilon_C = \frac{\sigma_C}{E} = \frac{144}{200000} \times 10^6 = 720\,\mu\varepsilon$$

$$k_a = \frac{\epsilon_C}{\epsilon_T} - 1 = \frac{720}{600} - 1 = 0.2$$

$$K = 1 + k_a k_b = 1 + 0.2 \times 0.8 = 1.16$$

From 4.3.1.2.5, flexural capacity calculations, $RF_{inventory}$ is 0.8, therefore:

$$RF_T = K \times RF_C = 1.16 \times 0.8 = 1.03$$

Hence, Neka's cross beams have adequate flexural capacity at the inventory level. RF_T is 1.19 for the operating level.

4.3.2.3 Axial Assessment: Horizontal Compressive Members

4.3.2.3.1 Fatigue Resistance

As discussed in Section 4.3, the maximum design live load for the bridge is 2963 kN, which includes a dynamic load factor of 1.1. During field load testing, however, the highest load recorded was 2146 kN, from a train configuration consisting of one locomotive and two loaded wagons. For this configuration, the ratio of the test load to the evaluation load, *T/W*, is 0.72. Assuming that the test results are eligible for extrapolation, k_b is set to 1 for this case. In contrast, for a lighter train configuration with a single locomotive only, T/W is 0.37. Accordingly, for this configuration, k_b is assigned a value of 0. Therefore, only the heavier train formation can be used for reassessing the load factors in axial members.

From Section 4.3.1.3 fatigue resistance calculations, RF was calculated to be 0.80. As shown in Figure 3.49, during dynamic load testing, the maximum strain observed in the top chord at the mid-span under a heavy train formation was –468 μstrain, which corresponds to a compressive stress of 93.6 MPa. The remaining fatigue life is then calculated using Miner's rule. Detail category C is considered for this member. The number of fatigue cycles is determined through the rainflow counting method, based on strain measurements, with a strain range of 468 μ strain:

$$\sigma_{T1} = \mathrm{E} \times \sigma_{T1} = 2 \times 10^5 \times 468 \times 10^{-6} = 93.6 \text{ MPa}$$

$$n_1 = 365 \times 25 \times 1 \times (20) = 1{,}825{,}000$$

$$(\Delta F)_n = \left(\frac{14.4 \times 10^{11}}{N}\right)^{\frac{1}{3}}$$

$$93.6 = \left(\frac{14.4 \times 10^{11}}{N}\right)^{\frac{1}{3}} \Rightarrow N_1 = 1.756 \times 10^6$$

$$R = 1 - D = 1 - \left(\frac{n_1}{N_1}\right) = 1 - \frac{0.1825 \times 10^6}{1.756 \times 10^6} = 0.90$$

An R value of 0.90 suggests that these members still have a significant fatigue life and are not concerning.

4.3.2.3.2 Axial Compressive Capacity

The maximum axial load in top chord member was 2816 kN for critical positioning of heavy train formation, as presented in Figure 4.13. Therefore, the axial capacity is:

$$\sigma_C = \frac{F}{A} = \frac{2816x10^3}{2 \times 14030} = 100.4 \,\text{MPa}$$

in which F is the axial load in the member, and A is its cross-sectional area. Therefore:

$$\varepsilon_C = \frac{\sigma_C}{E} = \frac{100.4}{200000} \times 10^6 = 502\,\mu\varepsilon$$

$$k_a = \frac{\epsilon_C}{\epsilon_T} - 1 = \frac{502}{468} - 1 = 0.07$$

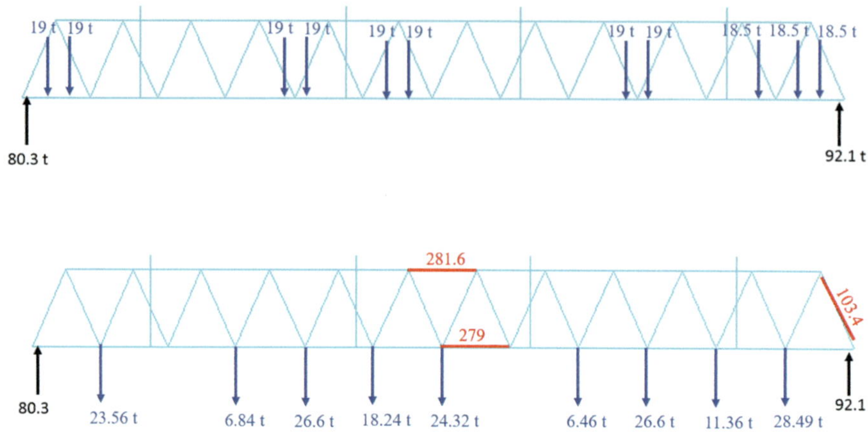

FIGURE 4.13 Loads in the truss members and nodes under train load.

$$K = 1 + k_a k_b = 1 + 0.07 \times 1 = 1.07$$

From 4.3.1.3 axial capacity calculations, we have $RF_{inventory}$ is 0.47, therefore:

$$RF_T = K \times RF_C = 1.07 \times 0.47 = 0.5$$

Given this is less than 1, it is repeated for operating level. $RF_{operating}$ is 0.61, therefore:

$$RF_T = K \times RF_C = 1.07 \times 0.61 = 0.65$$

Hence, Neka's horizontal members have inadequate axial compressive capacity in both inventory and operating levels.

4.3.2.4 Axial Assessment: Horizontal Tensile Members

4.3.2.4.1 Fatigue Resistance

From Section 4.3.1.4 fatigue resistance calculations, the RF was calculated to be 0.81. As shown in Figure 3.50, during dynamic load testing, the maximum strain observed in the bottom chord at the mid-span under a heavy train formation was 375 μstrain, which corresponds to a compressive stress of 75 MPa. The remaining fatigue life is then calculated using Miner's rule. The number of fatigue cycles is determined through the rainflow counting method, based on strain measurements, with a strain range of 375 μstrain:

$$\sigma_{T1} = E \times \sigma_{T1} = 2 \times 10^5 \times 375 \times 10^{(-6)} = 75\,\text{MPa}$$

$$n = 365 \times 25 \times 1 \times (20) = 1{,}825{,}000$$

$$(\Delta F)_n = \left(\frac{14.4 \times 10^{11}}{N}\right)^{\frac{1}{3}}$$

$$75 = \left(\frac{14.4 \times 10^{11}}{N}\right)^{\frac{1}{3}} \Rightarrow N_1 = 3.41 \times 10^6$$

$$R = 1 - D = 1 - \left(\frac{n_1}{N_1}\right) = 1 - \frac{0.1825 \times 10^6}{3.41 \times 10^6} = 0.95$$

Therefore, this member has no issues in terms of fatigue, given it has significant fatigue life remaining.

4.3.2.4.2 Axial Tensile Capacity

The maximum axial load in bottom chord member was 2790 kN for critical positioning of heavy train formation, as presented in Figure 4.12. Therefore, the axial capacity is:

$$\sigma_C = \frac{F}{A} = \frac{2790 \times 10^3}{2 \times 14030} = 99.4\,\text{MPa}$$

$$\varepsilon_C = \frac{\sigma_C}{E} = \frac{99.4}{200000} \times 10^6 = 497\,\mu\varepsilon$$

$$k_a = \frac{\epsilon_C}{\epsilon_T} - 1 = \frac{497}{375} - 1 = 0.33$$

$$K = 1 + k_a k_b = 1 + 0.33 \times 1 = 1.33$$

From Section 4.3.1.4 axial capacity calculations, we have $\text{RF}_{\text{inventory}}$ is 0.63, therefore:

$$\text{RF}_T = K \times \text{RF}_C = 1.33 \times 0.63 = 0.84$$

Given this is less than 1, it is repeated for the operating level. $\text{RF}_{\text{operating}}$ is 0.82, therefore:

$$\text{RF}_T = K \times \text{RF}_C = 1.33 \times 0.82 = 1.09$$

Hence, Neka's horizontal members have inadequate axial tensile capacity at the inventory level but are adequate at the operating level.

4.3.2.5 Axial Assessment: Diagonal Compressive Members

4.3.2.5.1 Fatigue Resistance

From Section 4.3.1.5 fatigue resistance calculations, RF was 0.75. As shown in Figure 3.51, during dynamic load testing, the maximum strain in the diagonal beam of the twelfth span of the deck reached 389 μstrain, corresponding to a stress of 77.8 MPa. The remaining fatigue life is then calculated using Miner's rule. Detail category A is considered for this member. The number of fatigue cycles is determined using the rainflow counting method, based on strain measurements. A range of 77.8 MPa corresponds to one cycle for each train passing, and 15 MPa corresponds to one cycle for each bogie passing:

$$\sigma_{T1} = E \times \sigma_{T1} = 2 \times 10^5 \times 389 \times 10^{(-6)} = 77.8\, MPa$$

$$\sigma_{T2} = E \times \sigma_{T2} = 2 \times 10^5 \times 7510^{(-6)} = 15\, MPa$$

$$n_{T1} = 365 \times 25 \times 1 \times (20) = 1{,}825{,}000$$

$$n_{T2} = 365 \times 25 \times 2 \times (20 \times 10) = 3{,}650{,}000$$

$$(\Delta F)_n = \left(\frac{82 \times 10^{11}}{N} \right)^{\frac{1}{3}}$$

$$77.8 = \left(\frac{14.4 \times 10^{11}}{N} \right)^{\frac{1}{3}} \Rightarrow N_1 = 17.41 \times 10^6$$

$$15 = \left(\frac{14.4 \times 10^{11}}{N} \right)^{\frac{1}{3}} \Rightarrow N_2 = 2.43 \times 10^9$$

$$R = 1 - D = 1 - \left(\frac{n_1}{N_1} + \frac{n_2}{N_2} \right) = 1 - \left(\frac{0.1825 \times 10^6}{17.41 \times 10^6} + \frac{3.65 \times 10^6}{2.43 \times 10^9} \right) = 0.99$$

Therefore, these members have no fatigue issues.

4.3.2.5.2 Axial Compressive Capacity

The diagonal members have already passed the compressive capacity check for the design load. However, for the sake of completeness, they are included here as well. The maximum axial load in the diagonal member was 1034 kN for critical positioning of heavy train formation, as presented in Figure 4.12. Therefore, the axial capacity is:

$$\sigma_C = \frac{F}{A} = \frac{1034 \times 10^3}{2 \times 6430} = 80.4MPa$$

$$\varepsilon_C = \frac{\sigma_C}{E} = \frac{80.4}{200000} \times 10^6 = 402\,\mu\varepsilon$$

$$k_a = \frac{\epsilon_C}{\epsilon_T} - 1 = \frac{402}{389} - 1 = 0.03$$

$$K = 1 + k_a k_b = 1 + 0.03 \times 1 = 1.03$$

From 4.3.1.5.2 axial compressive capacity calculations, $RF_{inventory}$ is 1.04, therefore:

$$RF_T = K \times RF_C = 1.03 \times 1.04 = 1.07$$

As expected, these members also pass the axial capacity under operational train loads.

4.3.2.6 Axial Assessment: Diagonal Tensile Members

This is not possible to perform, given that no measurement was performed for strain on this member.

4.3.2.7 Summary of Load Rating with Results of Load Tests

Tables 4.6 and 4.7 summarize the load rating factors after refining them using field load testing of Neka bridge. No updates are given for shear as it was not measured during the test campaign. It is evident that with the refined method, bending at the inventory level is now acceptable for both longitudinal and cross beams, which were deemed inadequate under design load. However, axial capacity of the top and bottom chord members is not adequate.

TABLE 4.6
Summary of load rating factors for flexural members of Neka Bridge

Element	Fatigue	Bending inventory	Bending operating
Longitudinal beam	remained life 96%	1.1	N.A.
Cross beam	remained life 22%	1.16	N.A.

TABLE 4.7
Summary of load rating factors for axial members of Neka Bridge

Element	Fatigue	Compressive/tensile (inventory)	Compressive/tensile (operating)
Top chord	remained life 90%	0.5	0.65
bottom chord	remained life 95%	0.84	1.09
Diagonal comp.	remained life 99%	1.07	N.A.

4.4 SUMMARY

This chapter presented the fundamentals of bridge load rating, detailing calculations from various methods and including a step-by-step example for the Neka case study. In this case study, the load rating was initially conducted using design loads; however, due to unsatisfactory results, it was further refined using data from field experiments.

Although the load rating improved for some members, others still demonstrated inadequate performance. In such cases, asset managers and chief engineers must determine appropriate actions, which may involve modifying the operational regime or implementing remedial and strengthening measures. The concepts and examples presented should aid readers in performing similar load rating assessments for their bridge assets.

REFERENCES

[1] American Association of State Highway and Transportation Officials, "AASHTO LRFD Bridge Design Specifications, 9th Edition," 2020.

[2] "Publication 139: Bridge Loading Code.".

[3] American Association of State Highway and Transportation Officials, "AASHTO Manual for Bridge Evaluation, 3rd Edition, with 2024 Interim Revisions," 2018 (with 2024 updates).

[4] International Union of Railways, "UIC Code 776-2R—Design Requirements for Rail-bridges Based on Interaction Phenomena between Train, Track and Bridge," 2009.

5 Frequency Analysis

5.1 MODAL ANALYSIS

Modal analysis of a structure is an approach used to describe the behaviour of a system by determining its natural frequencies, damping ratios, and mode shapes corresponding to each natural frequency. The simplest representation of a structure is a single-degree-of-freedom (SDOF) model, characterized by a mass, a spring, and a damper (Figure 5.1).

Two important parameters in modal analysis are natural frequency and damping ratio. The natural frequency is the frequency at which a system oscillates freely without the influence of external forces or damping. It is related to the stiffness (k) and mass (m) of the system and is expressed as:

$$f_n = \frac{\omega_n}{2\pi} = \frac{1}{2\pi}\sqrt{\frac{k}{m}}$$

where ω_n is the natural angular frequency.

- Stiffness (k): A higher stiffness results in a higher natural frequency, as the system resists deformation more strongly.
- Mass (m): A larger mass results in a lower natural frequency, as the system's inertia slows its oscillatory motion.

Natural frequency is crucial for understanding how a structure responds to dynamic loads. If an external force, such as wind, seismic activity, or traffic vibrations, matches the natural frequency, the structure may enter a state of resonance. This amplifies vibrations and can lead to structural damage or failure.

The damping ratio quantifies the system's ability to dissipate energy and describes how quickly oscillations diminish after an initial disturbance. It is usually given by:

$$\zeta = \frac{c}{2\sqrt{k.m}}$$

DOI: 10.1201/9781003494652-5

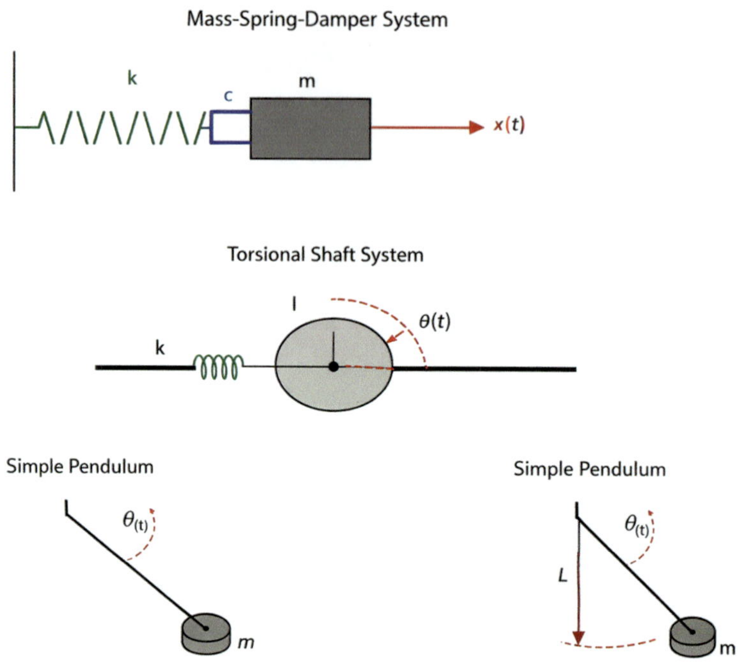

FIGURE 5.1 Dynamic model of a single-degree-of-freedom system (mass-spring-damper, torsional shaft, simple pendulum).

where *c* is the damping coefficient, which measures the system's resistance to motion due to damping forces.

The damping ratio determines the system's dynamic behaviour:

1. Underdamped ($\zeta < 1$): The system oscillates with decreasing amplitude.
2. Critically damped ($\zeta=1$): The system returns to equilibrium as quickly as possible without oscillating.
3. Overdamped ($\zeta>1$): The system returns to equilibrium slowly without oscillations.

Damping plays a crucial role in engineering design, as insufficient damping can result in prolonged vibrations, while excessive damping can reduce the system's efficiency or responsiveness.

As an example, consider the SDOF system of Neka bridge with the following parameters, as shown in Figure 5.2:

- Mass, m=11,340 kg,
- Stiffness, $k=1.2\text{x}10^7$ N/m,
- Damping coefficient, c=10000 N.s/m

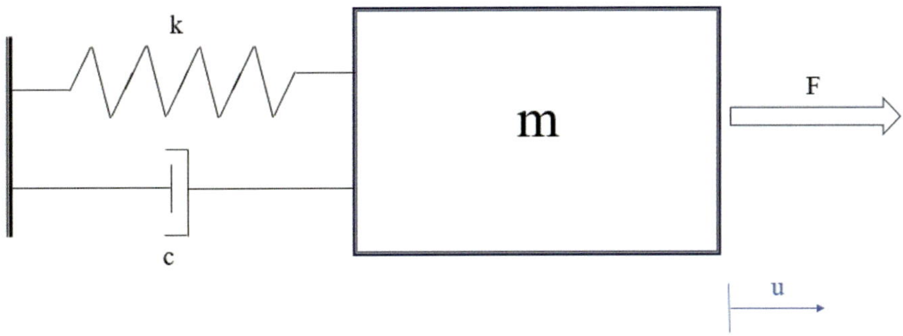

FIGURE 5.2 Dynamic model of an SDOF system.

Determine the natural frequency (f_n) and the damping ratio (ζ).

$$m\ddot{u} + c\dot{u} + ku = F$$

$$f_n = \frac{1}{2\pi}\sqrt{\frac{k}{m}} = \frac{1}{2\pi}\sqrt{\frac{1.2\times10^7}{11340}} =\approx 5.2\,\text{Hz}$$

$$\zeta = \frac{c}{2\sqrt{k.m}} = 0.013$$

The system has a natural frequency of approximately 5.2 Hz and a damping ratio of 0.013, indicating underdamped behaviour.

For the two DOF models of Neka bridge, we can determine the natural angular frequencies (ω_r) and the corresponding mode shapes (ψ_r) using the mass matrix [M] and stiffness matrix [K] provided below:

- Mass: $M = \begin{bmatrix} 11340 & 0 \\ 0 & 11340 \end{bmatrix}$ kg ,
- Stiffness: $K = \begin{bmatrix} 2.05 & 0.19 \\ 0.19 & 0.12 \end{bmatrix} \times 10^8 \ \frac{\text{N}}{\text{m}}$.

The equation of motion for this system is given by:

$$[M]\left\{\ddot{u}\right\} + [K]\{u\} = \{f\}$$

For free vibration (*f*=0):

$$[M]\{\ddot{u}\}+[K]\{u\}=\{0\}$$

Assume a harmonic solution:

$$\{u\}=\{\psi_r\}e^{-i\omega_r t}$$

Substituting into the equation:

$$([K]\{\psi_r\}-\omega_r^2[M]\{\psi_r\})e^{-i\omega_r t}=\{0\}$$

This simplifies to the eigenvalue problem:

$$[K]\{\psi_r\}=\omega_r^2[M]\{\psi_r\}$$

For a system with *n* degrees of freedom:

1. There are *n* natural frequencies $(\omega_1, \omega_2, \ldots, \omega_n)$. In this example *n*=2.
2. Each natural frequency has a corresponding mode shape $(\psi_1, \psi_2, \ldots, \psi_n)$.

The eigenvalue problem provides both the eigenvalues (ω_r^2) and eigenvectors (ψ_r):

$$u=\sum_{r=1}^{n}\psi_r e^{-i\omega_r t}$$

Mode shapes corresponding to different natural frequencies are orthogonal with respect to the mass and stiffness matrices:

$$\psi_s^T[M]\psi_r=0 \text{ and } \psi_s^T[K]\psi_r=0 \text{ for } \omega_r\neq\omega_s$$

This is proven by multiplying the eigenvalue equation for ω_s and ω_r by ψ_r^T and ψ_s^T, respectively, and subtracting the results:

$$0=\left(\omega_r^2-\omega_s^2\right)\psi_s^T[\mathrm{M}]\psi_\mathrm{r}$$

Since $\omega_r^2 \neq \omega_s^2$, it follows that:

$$\psi_s^T [M] \psi_r = 0 \text{ and } \psi_s^T [K] \psi_r = 0$$

To solve for the natural angular frequencies and mode shapes, the eigenvalue problem is solved numerically. Using MATLAB, the eigenvalues and eigenvectors are obtained as follows:

[*V*,*D*]=eig(*K*,*M*);

- Eigenvectors ([*V*]):

$$[V] = \begin{bmatrix} 0.93 & -0.09 \\ 0.09 & 0.93 \end{bmatrix}$$

- Eigenvalues ([*D*]):

$$[D] = \begin{bmatrix} \omega_1^2 & 0 \\ 0 & \omega_2^2 \end{bmatrix} = \begin{bmatrix} 18241 & 0 \\ 0 & 895 \end{bmatrix}$$

From [*D*], the natural angular frequencies are as follows. These represent the frequencies at which the system oscillates when excited, without external damping or forcing.

$$\omega_1 = \sqrt{18259} = 135.1 \frac{\text{rad}}{\text{s}} = 21.5\,\text{Hz}$$

$$\omega_2 = \sqrt{937} = 29.9 \frac{\text{rad}}{\text{s}} = 4.8\,\text{Hz}$$

The corresponding mode shapes are given by the columns of [*V*] as calculated below:

$$\psi_1 = \begin{bmatrix} 0.93 \\ 0.09 \end{bmatrix}, \psi_2 = \begin{bmatrix} -0.09 \\ +0.93 \end{bmatrix}$$

These describe the relative motion of the masses at each natural frequency. For this system:

- At ω_1 : Both masses move in phase (same direction).
- At ω_2 : The masses move out of phase (opposite directions).

Now, consider the case where the stiffness values are modified in this example as follows:
The modified stiffness matrix is given by

$$[K]=\begin{bmatrix} 2.05 & 0.19 \\ 0.19 & 0.12\times 0.8 \end{bmatrix}\times 10^8 \ \frac{\text{N}}{\text{m}}$$

The mass matrix remains unchanged:

$$[M]=\begin{bmatrix} 1 & 0 \\ 0 & 1 \end{bmatrix}\times 11340 \text{ kg}$$

The eigenvalue problem $([K]-\omega^2[M])\,\psi=0$ is solved numerically using MATLAB function described before:

- Eigenvectors ([V]):

$$[V]=\begin{bmatrix} 0.93 & -0.09 \\ 0.09 & +0.93 \end{bmatrix}$$

- Eigenvalues ([D]):

$$[D]=\begin{bmatrix} \omega_1^2 & 0 \\ 0 & \omega_2^2 \end{bmatrix}=\begin{bmatrix} 18239 & 0 \\ 0 & 685 \end{bmatrix}$$

From the eigenvalues, the natural angular frequencies are:

$$\omega_1=\sqrt{18239}\approx 135.1\frac{\text{rad}}{\text{s}}=21.5\,\text{Hz}$$

$$\omega_2=\sqrt{685}=26.2\frac{\text{rad}}{\text{s}}=4.2\,\text{Hz}$$

When compared to the initial case, the frequencies are slightly altered.

We are tasked with determining the natural angular frequencies (ω_r) and the corresponding mode shapes (ψ_r) for the system under the following assumptions: The mass matrix [M] and stiffness matrix [K] are defined as:

- Mass: $M = \begin{bmatrix} 5670 & 0 & 0 \\ 0 & 5670 & 0 \\ 0 & 0 & 5670 \end{bmatrix} \text{kg}$,
- Stiffness: $K = \begin{bmatrix} 6.14 & -5.15 & 1.13 \\ -5.15 & 8.13 & -5.65 \\ 1.13 & -5.65 & 8.27 \end{bmatrix} \times 10^7 \dfrac{\text{N}}{\text{m}}$.

To solve for the natural angular frequencies and mode shapes, the eigenvalue problem is solved numerically. Using MATLAB, the eigenvalues and eigenvectors are obtained as follows:

$$[V,D]=\text{eig}(K,M);$$

- Eigenvectors ([V]):

$$[V] = \begin{bmatrix} 0.75 & -0.94 & 0.57 \\ 0.94 & 0.18 & -0.92 \\ 0.58 & 0.92 & 0.76 \end{bmatrix}$$

- Eigenvalues ([D]):

$$[D] = \begin{bmatrix} \omega_1^2 & 0 & 0 \\ 0 & \omega_2^2 & 0 \\ 0 & 0 & \omega_3^2 \end{bmatrix} = \begin{bmatrix} 970 & 0 & 0 \\ 0 & 10613 & 0 \\ 0 & 0 & 28170 \end{bmatrix}$$

From [D], the natural angular frequencies are as follows. These represent the frequencies at which the system oscillates when excited, without external damping or forcing.

$$\omega_1 = \sqrt{970} = 31.1 \frac{\text{rad}}{\text{s}} = 4.96\,\text{Hz},$$

$$\omega_2 = \sqrt{10613} = 103 \frac{\text{rad}}{\text{s}} = 16.4\ \text{Hz}$$

$$\omega_3 = \sqrt{28170} = 167.8\frac{\text{rad}}{\text{s}} = 26.71\,\text{Hz}$$

The corresponding mode shapes are given by the columns of $[V]$ are calculated as follows:

$$\psi_1 = \begin{bmatrix} 0.75 \\ 0.94 \\ 0.58 \end{bmatrix}, \ \psi_2 = \begin{bmatrix} -0.94 \\ 0.18 \\ 0.92 \end{bmatrix}, \ \psi_3 = \begin{bmatrix} 0.57 \\ -0.92 \\ 0.76 \end{bmatrix}$$

5.2 MODAL ANALYSIS OF TRUSS BRIDGES

In Section 4.2, the stiffness matrix for Neka truss bridge was developed by assuming axial elements. By aggregating the mass of elements at the two end nodes, the mass matrix of the bridge can also be estimated. With the stiffness and mass matrices available and neglecting damping, the modal characteristics, including natural frequencies and mode shapes, can be calculated.

The developed truss bridge model for Neka Bridge includes 50 DOFs, as shown in Figure 5.3. To simplify the analysis, the element masses are lumped at the nodes based on the contributions of the connected elements. Some of the lumped node masses are calculated as follows:

$$\text{m}_2 = \text{one diagonal mass + one horizontal chord mass}$$
$$\text{+ one transverse beam mass + one longitudinal beam mass}$$

$$m_2 = \frac{1.65}{2} + \frac{3.2}{2} + \frac{6.56}{2} + \frac{1.71}{2} + 2.1 = 8.66 \text{ kN}$$

$$m_4 = 1.65 + \frac{3.2}{2} = 3.25\,\text{kN}$$

$$\text{m}_6 = \text{One pedestrian (across 3 meters) + One rail (across 3 meters)+5/2 sleepers}$$

$$m_6 = 1.65 + 3.2 + \frac{6.56}{2} + 1.71 + 4.2 = 14.04\,\text{kN}$$

$$m_8 = 1.65 + 3.2 = 4.85\,\text{kN}$$

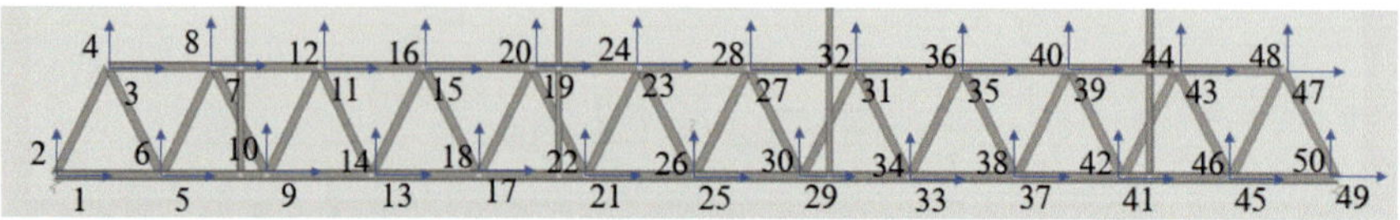

FIGURE 5.3 A 50 DOF model of Neka truss bridge.

The mass matrix is developed by assembling the lumped masses into a diagonal matrix, where each diagonal entry corresponds to the mass at a specific node, as shown below (only first 10x10 portion of mass matrix):

$$\begin{matrix} 866 & 0 & 0 & 0 & 0 & 0 & 0 & 0 & 0 & 0 \\ 0 & 866 & 0 & 0 & 0 & 0 & 0 & 0 & 0 & 0 \\ 0 & 0 & 325 & 0 & 0 & 0 & 0 & 0 & 0 & 0 \\ 0 & 0 & 0 & 325 & 0 & 0 & 0 & 0 & 0 & 0 \\ 0 & 0 & 0 & 0 & 1404 & 0 & 0 & 0 & 0 & 0 \\ 0 & 0 & 0 & 0 & 0 & 1404 & 0 & 0 & 0 & 0 \\ 0 & 0 & 0 & 0 & 0 & 0 & 485 & 0 & 0 & 0 \\ 0 & 0 & 0 & 0 & 0 & 0 & 0 & 485 & 0 & 0 \\ 0 & 0 & 0 & 0 & 0 & 0 & 0 & 0 & 1404 & 0 \\ 0 & 0 & 0 & 0 & 0 & 0 & 0 & 0 & 0 & 1404 \end{matrix}$$

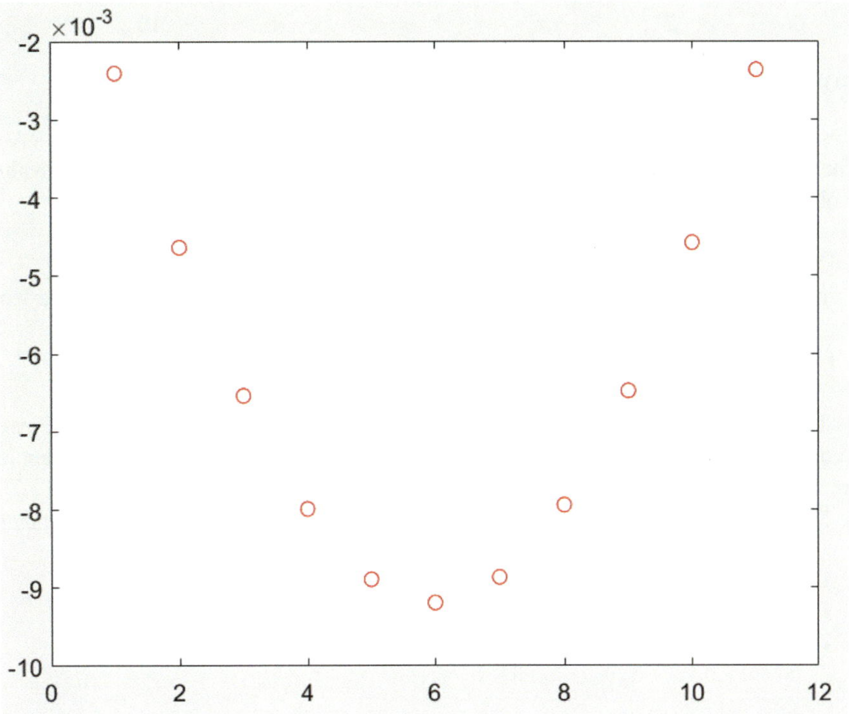

FIGURE 5.4 DOFs 6:6:46 for the first natural frequency of 4.9 Hz.

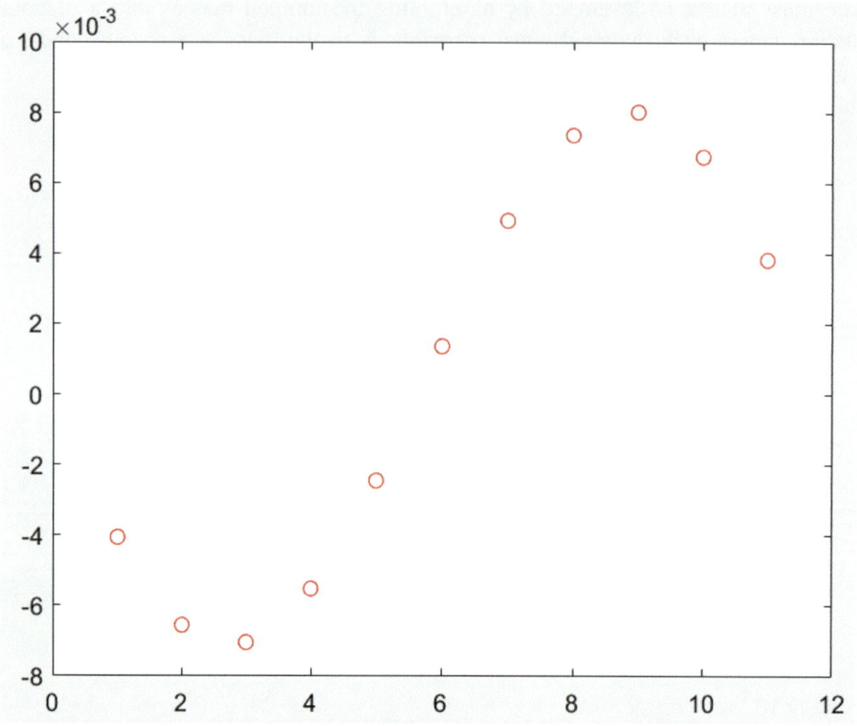

FIGURE 5.5 DOFs 6:6:46 for the second natural frequency of 15.6 Hz.

The modal characteristics are computed by solving the generalized eigenvalue problem described before using MATLAB, as shown in the function below:

[V, D] = eig(K(3:49, 3:49) × 9.81 × 100000, M(3:49, 3:49));
in which K(3:49, 3:49) applies the boundary conditions, 9.81 is for conversion from t/cm to kg/m, and 100000 is for conversion from kg to N.
f = sqrt(diag(D)) / (2 * pi); % Natural frequencies in Hz

The natural frequencies indicate the vibration characteristics of the bridge. Lower frequencies often correspond to global structural motions, while higher frequencies are associated with local element deformations.

Mode shapes reveal how the bridge deforms at each natural frequency. For instance:

- At 4.9 Hz, the bridge experiences the first mode shape (Figure 5.4).
- At 15.6 Hz, the second mode shape (Figure 5.5).
- At 19.9 Hz, the third mode shape (Figure 5.6).

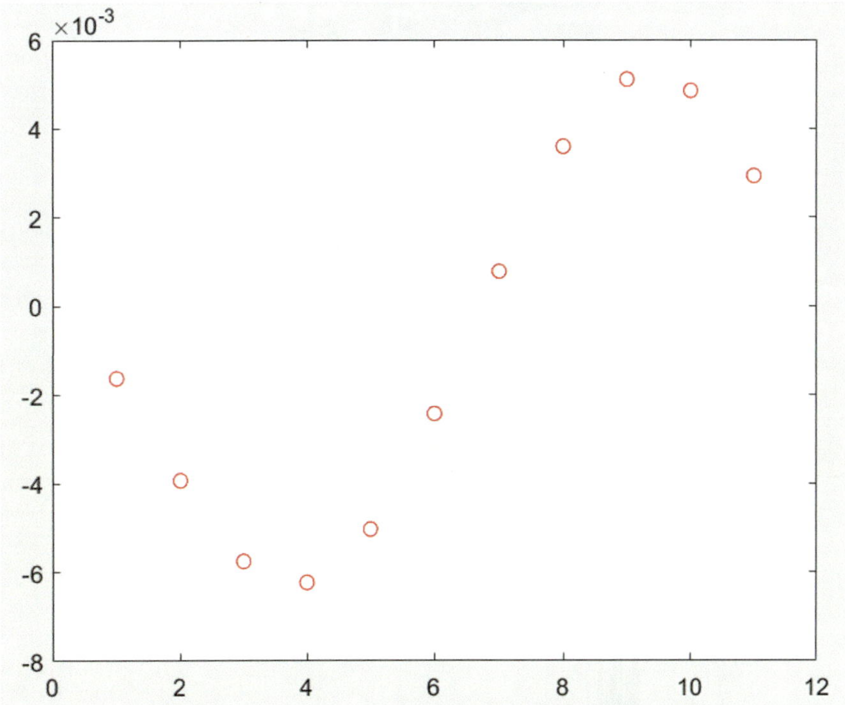

FIGURE 5.6 DOFs 6:6:46 for the third natural frequency of 19.9 Hz.

5.3 VIBRATIONAL TESTING OF STRUCTURES

5.3.1 Impact Testing

Impact testing is a widely used method in SHM and experimental modal analysis (EMA) to assess the dynamic properties of structures. It is particularly useful because it provides a quick and non-destructive way to excite the structure and gather critical information about its dynamic characteristics, such as natural frequencies, mode shapes, and damping ratios.

This method is not governed by a specific standard but is performed based on engineering judgement and experience. In this test, an instrumented hammer or a weight drop system is used to generate an impulse force (Figure 5.7), which excites the structure over a broad frequency range. The response of the structure is typically measured using accelerometers or laser vibrometers placed at multiple points. The force input and structural response are processed using fast Fourier transform (FFT) to compute the FRF, which provides insight into the system's dynamic behaviour.

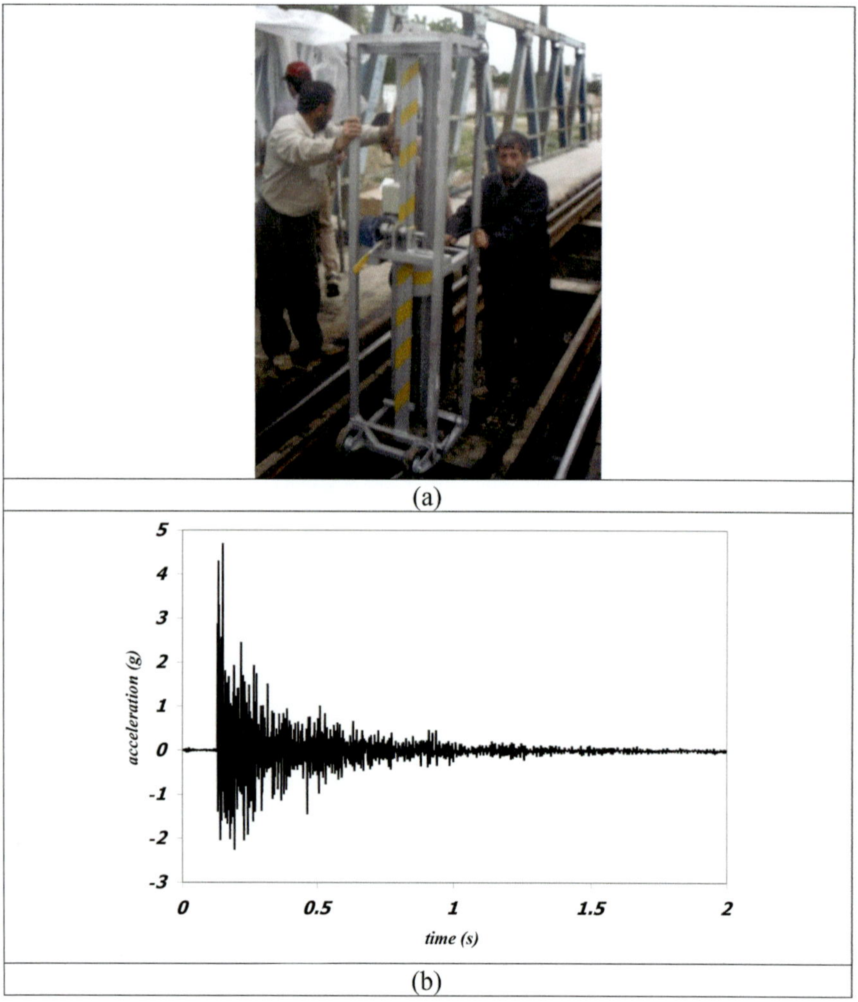

FIGURE 5.7 (a) Impact testing of Neka Bridge by dropping a 70-kg weight from a height of 1.65 m at the centre of each transverse beam and (b) Vertical acceleration at the mid-span of the southern truss corresponding to the release of a 70-kg weight from a height of 1.65 m at the centre of the bridge's middle transverse beam.

5.3.2 Shaker Testing

Another type of loading that can be used in field loading tests is *shaker testing*. Shaker testing, also known as force excitation testing, is a technique to assess the dynamic response of large structures under controlled loading conditions. This method is particularly useful in situations where the structure's natural dynamic behaviour needs to be excited and observed in a repeatable and controlled manner.

In shaker testing, a force generator is used to apply a controlled dynamic load to the structure. The load can be sinusoidal (varying in a regular, smooth pattern) or random (such as white noise or periodic noise). The structure's response is measured at different locations using accelerometers. The output data is analysed to derive the frequency response function (FRF), which provides insights into the structure's dynamic properties, such as its natural frequencies, mode shapes, and damping ratios.

The excitation signal can be either continuous or discrete, each with its unique characteristics:

- *Continuous signals*: These include pure random noise (which has a broad frequency spectrum) or periodic noise (which contains distinct frequency components). These signals are ideal for testing a structure's response over a wide frequency range.
- *Discrete signals*: These include impulsive random noise or chirp signals, which are designed to excite the structure at specific frequencies or over a limited frequency range. These signals are useful for pinpointing specific dynamic characteristics of a structure.

There are different types of shakers as well:

- *Electromagnetic shakers*: These shakers use electromagnetic forces to generate a sinusoidal or random force. They are commonly used for testing smaller or medium-sized structures where a high level of precision and control over the excitation force is required.
- *Electrohydraulic (Servo-hydraulic) shakers*: These are used for larger structures and can generate higher force outputs, making them suitable for civil engineering applications, including testing large-scale bridges or buildings. They are highly versatile, able to apply both sinusoidal and random loads.
- *Inertial shakers*: These shakers generate force using a rotating mass with an unbalanced weight, creating a dynamic force through inertia. This type of shaker is often used in situations where a continuous or reciprocating motion is needed.

There is no specific standard for conducting this test on civil engineering structures, and its execution requires experience and engineering judgement. A few samples are shown in Figure 5.8.

5.3.3 Ambient Vibration Testing

Ambient vibration testing is a passive method of measuring the dynamic response of a structure. It leverages the ambient forces that already exist in the environment, such as wind or vehicle traffic, which produce vibrational energy that can be measured. Unlike active methods (e.g., shaker or impact testing), there is no need to introduce an external force to excite the structure. This makes it an attractive option in scenarios where cost, time, or practicality of active testing methods are limiting factors.

(a)

(b)

FIGURE 5.8 Railway bridge shaker in (a) lateral and (b) vertical excitation modes.

In environmental vibration testing, the structure responds to the natural forces acting upon it, and the goal is to capture this response through precise measurements. Seismographs (also known as accelerometers or geophones) (Figure 5.9) are deployed on the structure at various points to measure the vibrations. These sensors record the time-series data of the vibrations, which is then analysed to derive the modal properties of the structure, including natural frequencies, mode shapes, and damping ratios.

5.3.4 Dynamic Testing

During dynamic load testing, which was discussed in depth in previous chapters, one or more specified vehicles cross over the structure (Figure 5.10). Typically, the heaviest allowed vehicle is used, and it is controlled to pass over the structure at specific speeds while the structure's responses are measured. In dynamic testing, the acceleration responses of the structure are of particular interest, and the free vibration response of the structure is analysed to extract its natural frequencies, damping, and mode shapes.

FIGURE 5.9 Three-component seismograph in the frequency range of 0.0167–50 Hz.

FIGURE 5.10 Measuring bridge response to crossing of two trains over 1035 km bridge.

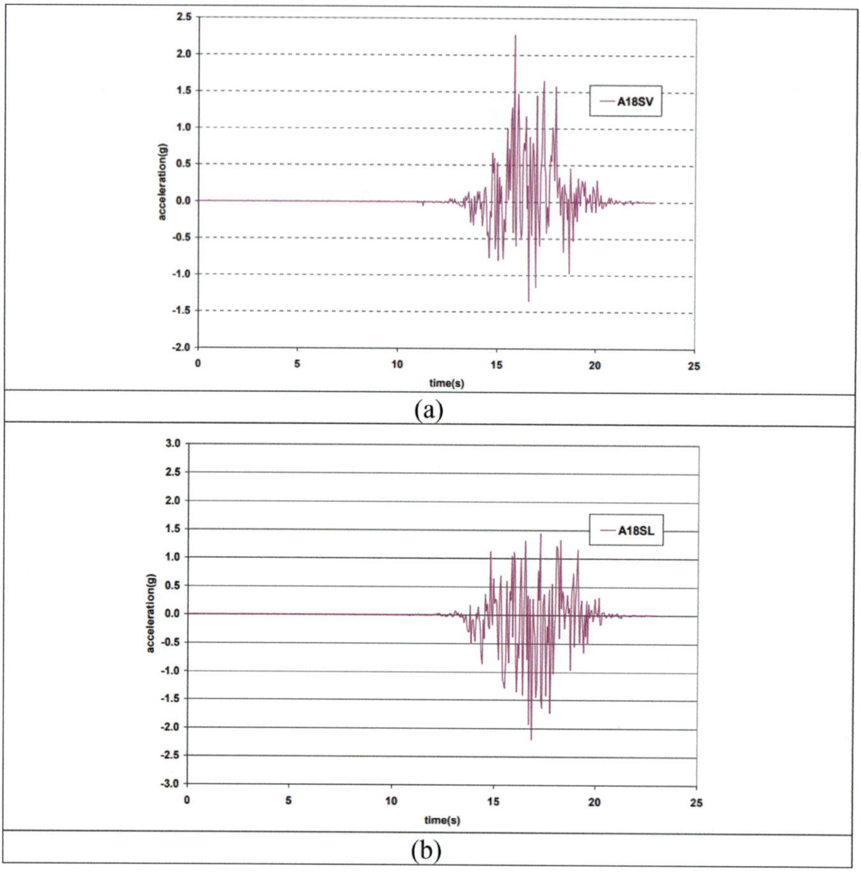

FIGURE 5.11 (a) Vertical and (b) lateral accelerations at the mid-span of the lower chord of the southern truss of Neka Bridge during the passage of a single diesel engine.

The acceleration at the mid-span in the dynamic test of the Neka truss bridge with the passage of a single diesel engine is presented in Figure 5.11. The maximum vertical and lateral accelerations at the mid-span of the lower chord of the southern truss reach 2.2g, and the maximum lateral acceleration at the mid-span of the upper chord of the southern truss reaches 1.8g. The maximum acceleration increases with the speed of the load passage.

5.4 EXPERIMENTAL MODAL ANALYSIS

During vibration testing, the response of the structure to dynamic excitation is commonly recorded through installed accelerometers on the structure. The raw acceleration data needs to be processed to extract meaningful information such as natural frequencies, damping ratios, and mode shapes.

There are various methods to achieve this. One common approach is to utilize the time-domain response of the structure, from which the dynamic properties can be inferred. However, time-domain methods can sometimes be less effective when dealing with complex signals that contain multiple overlapping frequencies. In such cases, the conversion of the measured acceleration into dynamic properties is more effectively carried out in the frequency domain.

In the frequency domain, the measured acceleration is transformed into a spectrum that reveals the different frequency components present in the vibration response. This transformation is crucial because the frequency domain provides a clearer picture of the structure's behaviour by isolating the natural frequencies, which are the key parameters in dynamic analysis. The main information in this representation is expressed as the coefficients of the similarity between the acceleration response and orthogonal sinusoidal (sine) and cosine signals, which are the basis functions of harmonic motion.

This transformation from the time domain to the frequency domain is typically achieved using FFT, which is a computationally efficient algorithm that converts time-domain data into its corresponding frequency spectrum. It breaks down the complex vibration signal into its constituent frequencies, allowing the natural frequencies of the structure to be identified.

The resulting frequency spectrum is plotted as amplitude versus frequency, with prominent peaks indicating the natural frequencies of the structure, as shown in Figure 5.12. These peaks correspond to the resonant frequencies at which the structure tends to vibrate with the greatest amplitude. Additionally, the phase information obtained from the FFT can be used to determine the relative displacement's sign between different points on the structure, which is key in identifying the mode shapes.

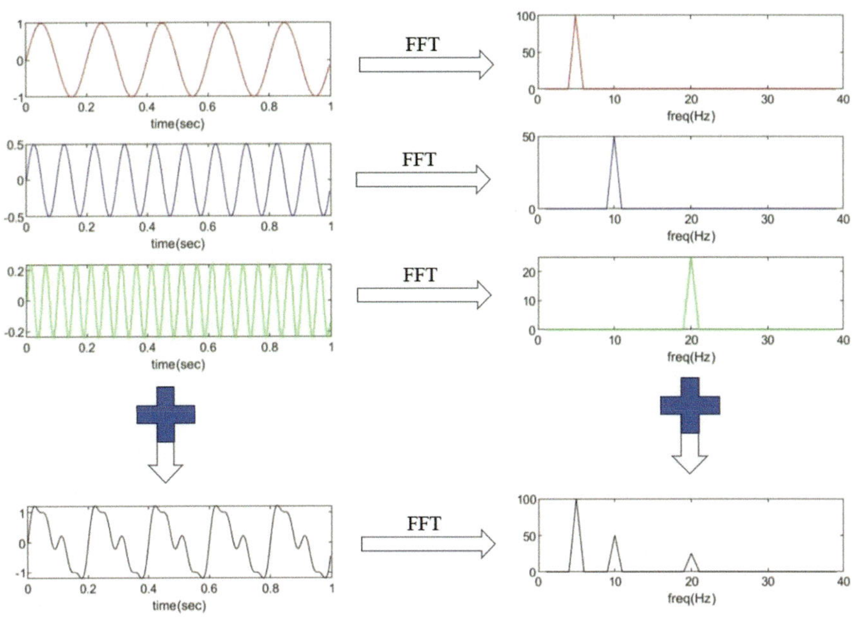

FIGURE 5.12 Fourier transform of the data with frequency base signals.

The physical concept of this is that the vibrational signal passes through a series of adjustable bandpass filters with multiple different central frequencies. For each frequency band, the root mean square (RMS) of the signal is measured using a voltmeter, and the RMS of each frequency band is plotted. If this is done on the free vibration signal of the structure, a high RMS value in a specific frequency range indicates the presence of the structure's natural frequency in that range.

5.4.1 Fourier Transform

Mathematically, the conversion of a time-domain signal to the frequency domain is performed using the Fourier transform. The Fourier transform provides a mathematical way to decompose a function (such as a time-varying signal) into its constituent frequencies, offering a powerful tool for analysing the frequency content of a signal.

The Fourier transform of a complex number at any frequency f is defined as the integral of the function $x(t)$, the time-domain signal, multiplied by the exponential function $e^{-i2\pi ft}$. The exponential function in this case represents a complex sinusoidal oscillation at frequency f. Mathematically, it is written as:

$$F\left[x(t)\right] = X(f) = \int_{-\infty}^{+\infty} x(t)e^{-i2\pi ft}\,\mathrm{d}t$$

This transform takes the time-domain signal $x(t)$ and represents it as a sum of sinusoidal components, each with a particular frequency and amplitude. The function $e^{-i2\pi ft}$ can be expressed as a combination of cosine and sine functions, where the cosine term represents the even (symmetric) component of the signal and the sine term represents the odd (asymmetric) component. Therefore, any arbitrary function $x(t)$ can be decomposed into even and odd functions.

The real part of the Fourier transform, $X(f)$, corresponds to the even function, representing the even or symmetric component of the signal. The imaginary part of $X(f)$ corresponds to the odd function, which represents the odd or asymmetric component of the signal. This decomposition allows for a clearer analysis of the signal's symmetrical and asymmetrical properties.

$$X(f) = \int_{-\infty}^{+\infty} x(t)\cos(2\pi ft)\,\mathrm{d}t - i\int_{-\infty}^{+\infty} x(t)\sin(2\pi ft)\,\mathrm{d}t$$

In this expression, the first integral represents the even part of the signal, while the second integral represents the odd part. The cosine function is an even function, and the sine function is an odd function. By using both, the Fourier transform breaks the signal $x(t)$ into its respective even and odd components.

To further illustrate this, let's define the even and odd components of the signal $x(t)$ as follows:

$$x_e(t) = \frac{1}{2}\left[x(t) + x(-t)\right]$$

$$x_0(t) = \frac{1}{2}\left[x(t) - x(-t)\right]$$

where $x_e(t)$ represents the even component, and $x_0(t)$ represents the odd component of the function $x(t)$. This decomposition shows that any signal can be viewed as a combination of even and odd components, which is particularly useful in signal analysis (Figure 5.13).

Parseval's theorem provides a crucial relationship between the time-domain and frequency-domain representations of a signal. It states that the total energy of a signal in the time domain is equal to the total energy of its Fourier transform in the frequency domain. This can be expressed as:

$$\int_{-\infty}^{+\infty} x(t) \cdot y^*(t)\,\mathrm{d}t \rightarrow \int_{-\infty}^{+\infty} X(f) \cdot Y^*(f)\,\mathrm{d}f$$

where $y^*(t)$ is the complex conjugate of $y(t)$, and similarly for $Y^*(f)$. This theorem implies that the interaction (or inner product) between two functions in the time domain is equivalent to the interaction between their Fourier transforms in the frequency domain. It offers insights into how energy or power is distributed between time and frequency components.

When we replace $y(t)$ with $x(t)$, we obtain:

$$\int_{-\infty}^{+\infty} x(t) \cdot x^*(t)\,\mathrm{d}t \rightarrow \int_{-\infty}^{+\infty} X(f) \cdot X^*(f)\,\mathrm{d}f$$

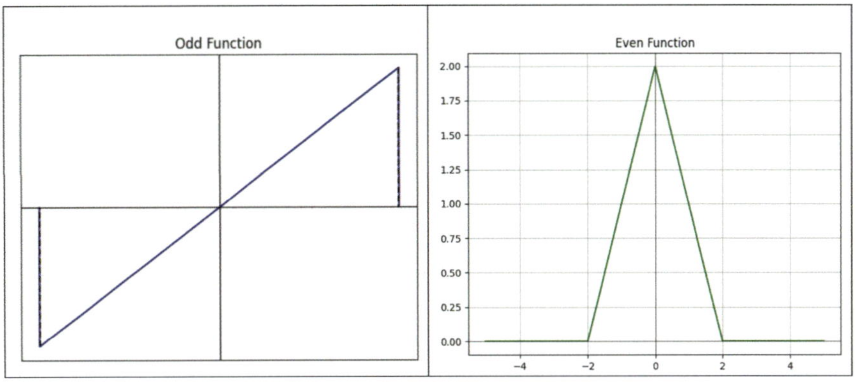

FIGURE 5.13 Any function can be represented as the sum of an even and odd function.

This equation reveals that the integral of the product of a function and its complex conjugate in the time domain is equal to the integral of the product of its Fourier transform and its complex conjugate in the frequency domain. In simpler terms, the total energy (or power) of a signal is conserved across both domains, ensuring that the frequency-domain representation carries the same amount of information as the time-domain representation.

Thus, Parseval's theorem essentially states that the sum of the squares of a function in the time domain is equivalent to the sum of the squares of its Fourier transform in the frequency domain. This principle is particularly useful in signal processing and structural health monitoring, where energy conservation plays a key role in understanding the behaviour of systems.

5.4.2 Discrete Fourier Transform

When analysing signals using computers, the Fourier transform must be discretized. The discrete Fourier transform (DFT) is applied to a signal vector $x(n)$ of length N at frequency k. The DFT is calculated by taking the dot product of the signal vector with the base signal $e^{-\frac{i2\pi kn}{N}}$, where the frequency k can range from 0 to $N-1$. The formula for the DFT is:

$$X(k)=\sum_{n=0}^{N-1}x(n)e^{-\frac{i2\pi kn}{N}}$$

Alternatively, the DFT can be expressed as:

$$X(k)=\sum_{n=0}^{N-1}x(n)\cos\left(\frac{2\pi kn}{N}\right)-i\sum_{n=0}^{N-1}x(n)\sin\left(\frac{2\pi kn}{N}\right)$$

In this expression:

- The cosine term represents the real part of the DFT, which is an even function.
- The sine term represents the imaginary part of the DFT, which is an odd function.

The DFT is periodic with a period of N, meaning that $X(k+N)=X(k)$. The frequency $k=0$ corresponds to the average or sum of the signal components over the entire length of the signal. The DFT at $k=\frac{N}{2}$ corresponds to the Nyquist frequency, which represents the highest frequency component in the signal. The value at $X\left(\frac{N}{2}\right)$ is a real number.

The DFT is also periodic in nature, so frequencies greater than *N*/2 represent negative frequencies. X(−k) = X(N−k)

$$X\left(\frac{N}{2}\right)=\sum_{n=0}^{N-1}x(n)e^{-\frac{i\pi n}{2}}=\sum_{n=0}^{N-1}x(n)(-1)^n$$

The frequency step Δf of the DFT is given by:

$$\Delta f=\frac{1}{N.\Delta t}=\frac{f_s}{N}$$

where f_s is the sampling frequency, and Δt is the sampling period.
The inverse DFT can be defined to reconstruct the signal from its frequency domain representation:

$$x(n)=(1/\mathrm{N})\sum_{k=0}^{N-1}X(k)e^{-\frac{i2\pi kn}{N}}$$

This allows the signal to be transformed back from the frequency domain to the time domain.

This mathematical framework is essential in many signal processing applications, such as analysing sound waves, images, and other types of data that can be represented in terms of their frequency components.

For example, let's calculate the discrete Fourier transform of $x(n)=\cos\left(\frac{4\pi n}{N}\right)$ for n = 0 to 7. The signal $x(n)$ is defined as a cosine wave:

$$x(n)=\cos\left(\frac{4\pi n}{N}\right)$$

where $n = 0$ to 7, and the length of the signal is N=8. This signal repeats twice within the block length, and the Fourier transform of the signal has non-zero values for frequencies at k=2 and k=6.

The corresponding DFT, $X(k)$ is computed for N = 8 elements. For the cosine signal $x(n)$, the DFT yields the following vector:

$$x=\{1\quad 0\quad -1\quad 0\quad 1\quad 0\quad -1\quad 0\}$$

And the DFT of this signal is:

$$X=\{0\quad 0\quad 4\quad 0\quad 0\quad 4\quad 0\quad 0\}$$

This indicates that the frequency components of the signal are concentrated at $k=2$ and $k=6$, with non-zero values, corresponding to the cosine signal's repetitions.

5.4.2.1 DFT of Neka Bridge Dynamic Test Results

For the Neka Bridge dynamic test, the signal data from measurement channels (e.g., Channel 5 and Channel 23) is loaded into MATLAB. The DFT is computed for each signal, focusing on the free vibration section of the signal after the diesel has passed.

1. Loading the Signal Data: The vertical acceleration data from Channel 5 and Channel 23 is loaded from the specified file paths:
   ```
   load('C:\CH05.ASC')
   load('C:\CH23.ASC')
   ```
2. Plotting the Free Vibration Section: The free vibration section of the signals from both channels is plotted (from data point 16335 to the end, as shown in Figure 5.14):
   ```
   plot(CH05(16335:end))
   hold on
   plot(CH23(16335:end))
   ```
3. DFT Calculation for Multiple Channels: The DFT is calculated for the free vibration sections of various channels (5, 8, 12, 14, 20, and 23):
   ```
   [ff05] = fft(CH05(16335:end));
   [ff08] = fft(CH08(16335:end));
   [ff12] = fft(CH12(16335:end));
   [ff14] = fft(CH14(16335:end));
   [ff20] = fft(CH20(16335:end));
   [ff23] = fft(CH23(16335:end));
   ```

   The frequency vector is also defined:
   ```
   fr = 0:2000/19666:2000-2000/19666;
   ```

   Here, $f_s = 2000\,\text{Hz}$ is the sampling frequency, and frequency steps are computed based on the signal length.
4. Plotting the magnitude of the DFT: The magnitude of the DFT for each channel is plotted to visualize the frequency content of the signals:
   ```
   figure(1);
   plot(fr(1:250), abs(ff05(1:250)), 'r-', fr(1:250), abs(ff08(1:250)), 'b-', ...
      fr(1:250), abs(ff12(1:250)), 'p-', fr(1:250), abs(ff14(1:250)), 'c-', ...
      fr(1:250), abs(ff20(1:250)), 'k-', fr(1:250), abs(ff23(1:250)), 'g-')
   legend('CH05', 'CH08', 'CH12', 'CH14', 'CH20', 'CH23')
   ```

In Figure 5.15, the frequency content of the bridge's free vibration in the vertical direction is presented for channels 5, 8, 12, 20, and 23, which measure the vertical acceleration of the southern truss at 6, 13, 18, 26, and 30 m from the support, as well as channel 14, which measures the vertical acceleration of the northern truss at 18 m (mid-span). Peaks are observed at frequencies of 2.3, 4.3, 5.1, 9.3, 12.2, and 17.6 Hz, with the peak at 4.3 Hz being the largest.

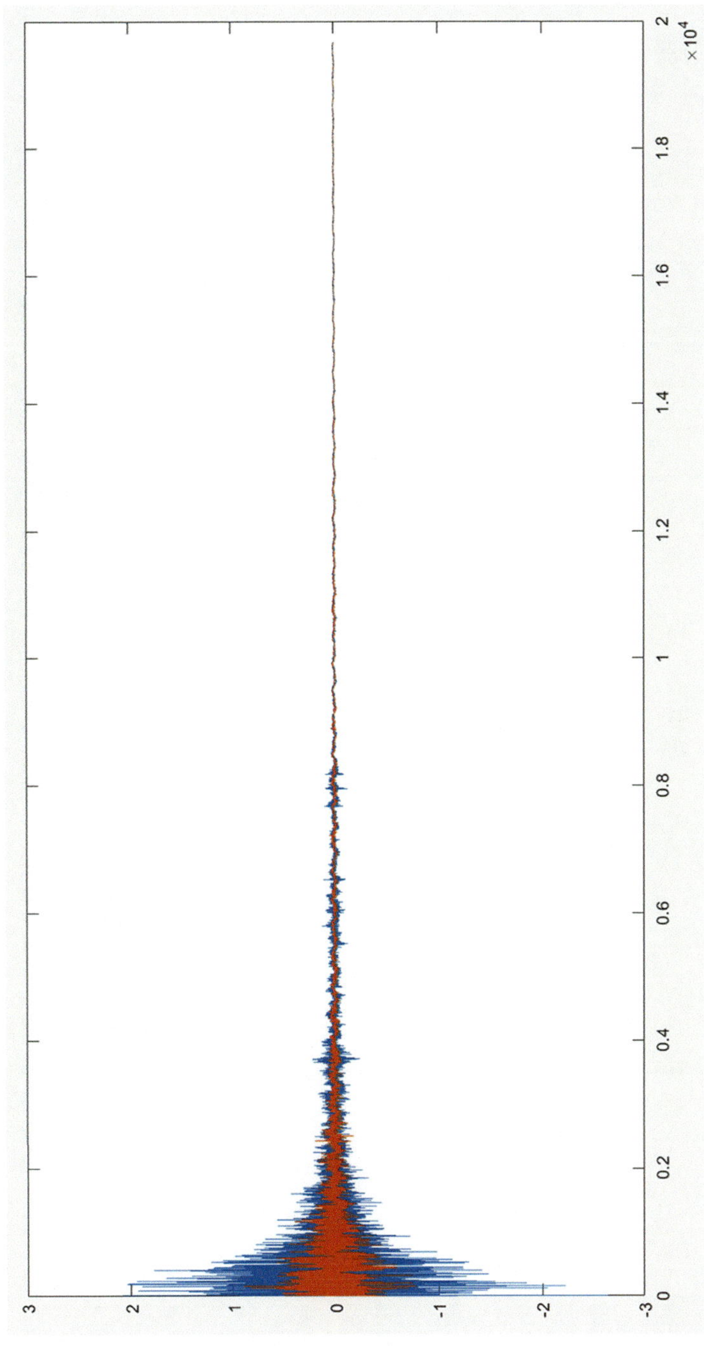

FIGURE 5.14 Free vibration section of the vertical acceleration signals from Channel 5 and Channel 23 after the diesel passes.

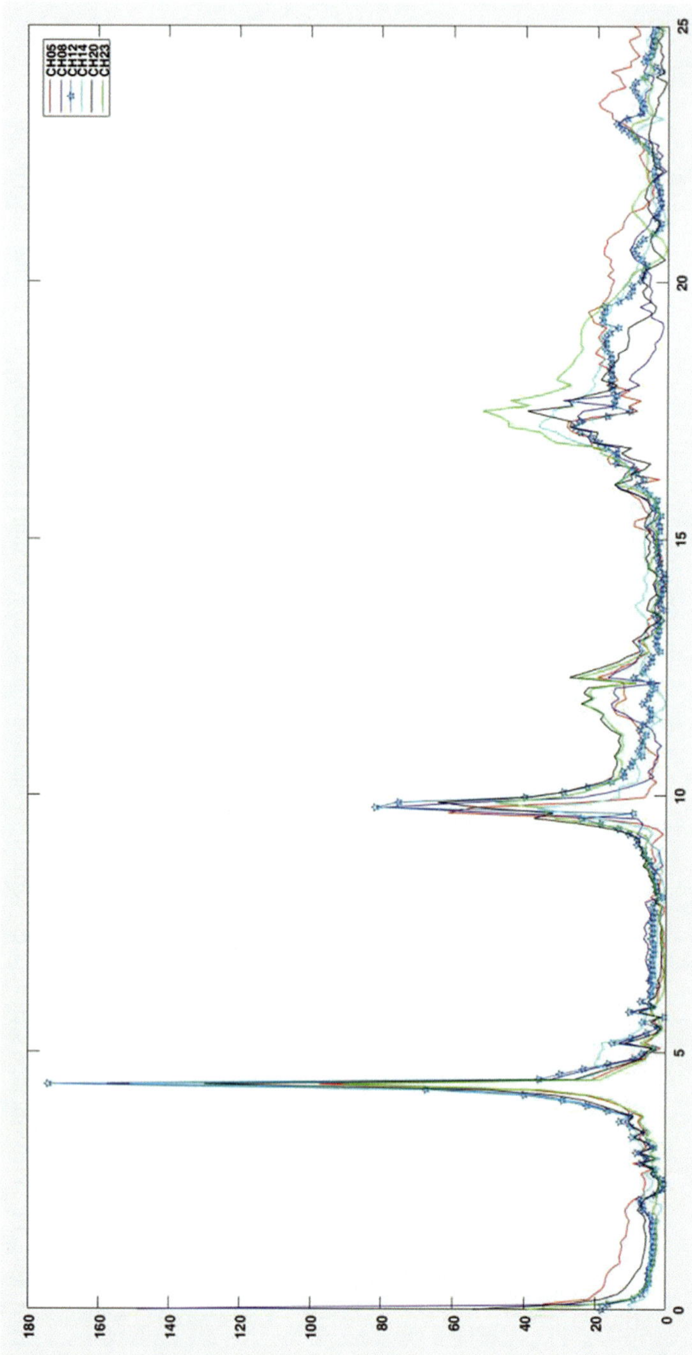

FIGURE 5.15 Magnitude of the DFT for the free vibration sections of five measurement channels after the diesel passes, showing the frequency components of each signal.

5.4.3 Spectral Density Function Overview

The spectral density function is a key concept in signal processing that describes how the power or energy of a signal is distributed over frequency. When dealing with measured signals, random noise can obscure the frequency content, so alternative techniques like the autocorrelation or cross-correlation functions can help mitigate this issue.

The autocorrelation function $R_{xx}(\tau)$ measures the similarity of a signal $x(t)$ with a time-shifted version of itself $x(t-\tau)$. It is mathematically defined as:

$$R_{xx}(\tau) = E\left[x(t).x(t-\tau)\right]$$

where $E[\cdot]$ represents the expectation operator, t is time, and τ is the time lag.

The spectral density function $S_{xx}(f)$ is obtained as the Fourier transform of the autocorrelation function:

$$F\left[R_{xx}(\tau)\right] = S_{xx}(f) = \int_{-\infty}^{+\infty} R_{xx}(\tau) e^{-i2\pi f\tau} \mathrm{d}\tau$$

This function shows how the signal's power is distributed across different frequencies f.

The cross-correlation function $R_{yx}(\tau)$ measures the similarity between two different signals $x(t)$ and $y(t)$, as one signal is shifted in time relative to the other:

$$R_{yx}(\tau) = E\left[y(t).x(t-\tau)\right]$$

The cross-spectral density function $S_{yx}(f)$ is the Fourier transform of the cross-correlation function:

$$F\left[R_{yx}(\tau)\right] = S_{yx}(f) = \int_{-\infty}^{+\infty} R_{yx}(\tau) e^{-i2\pi f\tau} \mathrm{d}\tau$$

This function describes how the signals $x(t)$ and $y(t)$ are related in the frequency domain, capturing any shared frequency components between them.

When a signal contains random noise:

- The *autocorrelation function* helps suppress the noise by averaging over time lags τ, which enhances the signal's periodic or deterministic features.
- Computing the spectral density function using the Fourier transform of the autocorrelation function further identifies the key frequency components, isolating them from noise.

Similarly, in the case of cross-spectral analysis between two signals, this approach helps identify correlated frequency content, even in the presence of noise.

5.4.4 Frequency Leakage and Windowing

In DFT, it is assumed that the block of data being transformed is periodic and repeats indefinitely. If the chosen block's endpoints are not aligned with the periodicity of the actual signal, additional frequency content can appear in the Fourier transform. This phenomenon is known as frequency leakage, where energy from the main frequencies leaks into other frequencies. Figure 5.16 illustrates frequency leakage, which occurs when the selected block's repetition is not equivalent to the entire measured signal.

To mitigate the effects of frequency leakage, one can apply a windowing function to the data block prior to performing the Fourier transform. A windowing function smooths the edges of the block by tapering the values toward zero at both ends. This reduces discontinuities at the boundaries and minimizes the introduction of artificial frequencies during the Fourier transform. Figure 5.17 shows the effect of windowing, where the impact of the block's endpoints is reduced.

Various windowing functions are commonly used, each with distinct characteristics, a few of which are presented in Figure 5.18. For example, the Hann window provides a smooth tapering shape and is suitable for general spectral analysis. The Hamming window, which is similar in shape to the Hann window, reduces side lobe levels more effectively. The Blackman window offers even a greater suppression of side lobes but at the expense of reduced frequency resolution. In contrast, the rectangular window, equivalent to applying no windowing at all, often leads to severe frequency leakage. Figure 5.19 shows the impacts of windowing.

The following steps summarize the process of applying windowing and estimating the spectral density using smaller segments of the signal:

1. Divide the Signal:
   - Split the stationary signal into smaller overlapping or non-overlapping segments. Overlapping segments reduce data loss at the block boundaries.
   - Example: For a signal with N=36,000 elements, you could divide it into blocks of N_b=2048 samples with 50% overlap.
2. Apply Windowing:
   - Apply a window function (e.g., Hann, Hamming, or Blackman) to each segment. This step tapers the endpoints and minimizes discontinuities.
3. Compute the Fourier Transform:
   - Perform the Fourier transform for each windowed segment to extract its frequency content.
4. Average the Results:
   - Average the Fourier transforms (or their squared magnitudes) of all segments to reduce random noise and obtain a smoothed spectral density estimate.

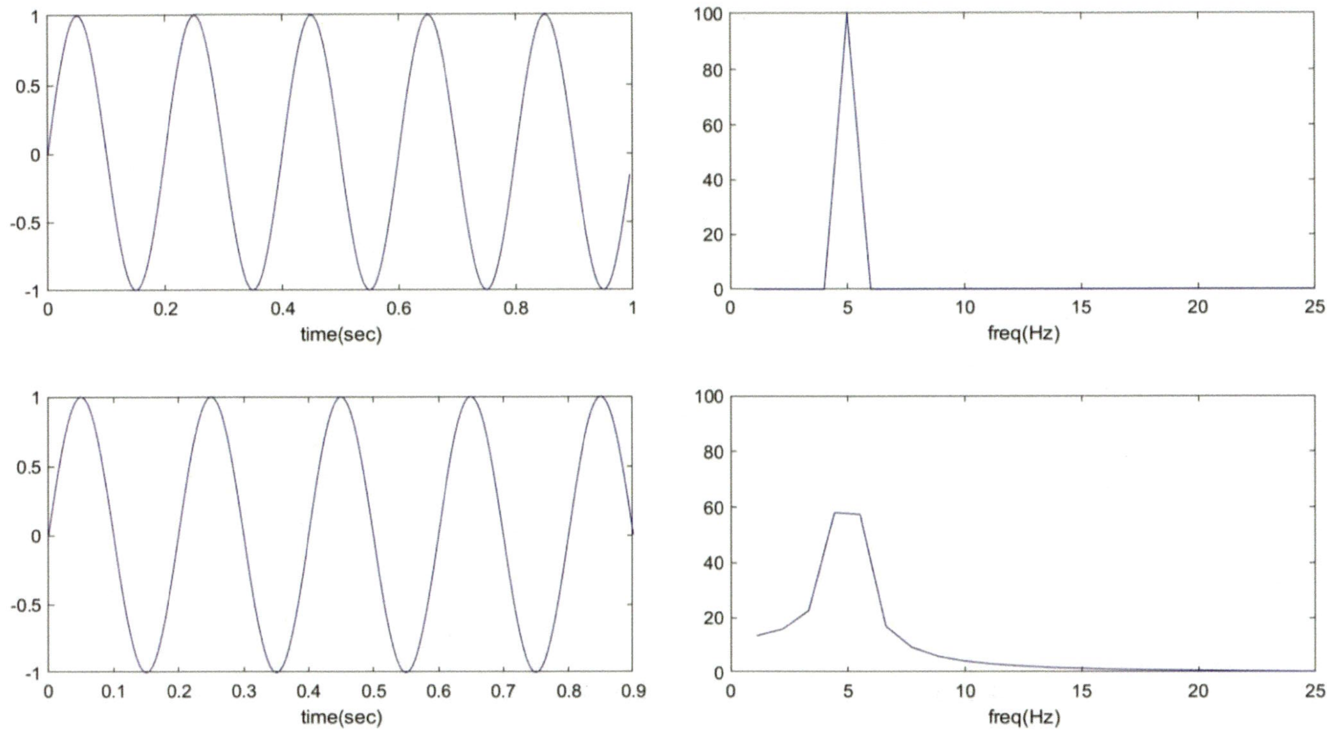

FIGURE 5.16 Frequency leakage when the selected block's repetition is not equivalent to the entire measured signal.

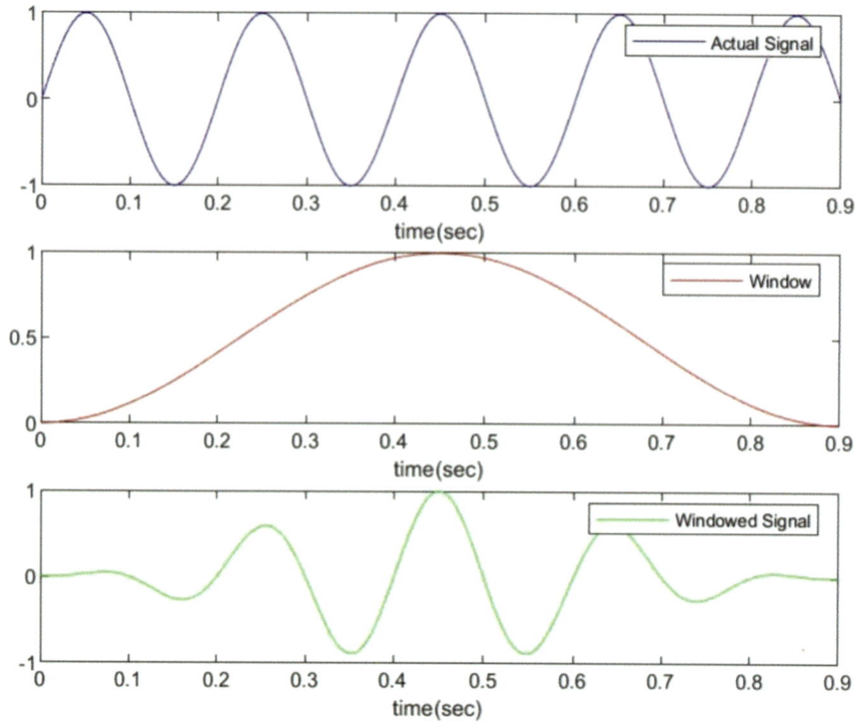

FIGURE 5.17 Effect of windowing on reducing block's endpoints.

The following MATLAB code applies windowing, computes the Fourier transform, and averages the results:

```
% Example: Signal with N = 36000 samples, divided into smaller segments
fs = 2000;              % Sampling frequency in Hz
signal = randn(1, 36000); % Replace with your actual signal
N_b = 2048;             % Block size
overlap = 1024;         % Overlap size (50% of block size)
window = hann(N_b);     % Hann window
nfft = N_b;             % Number of FFT points

% Calculate the spectral density
[S, f] = pwelch(signal, window, overlap, nfft, fs);

% Plot the spectral density
figure;
plot(f, 10*log10(S));
xlabel('Frequency (Hz)');
ylabel('Power/Frequency (dB/Hz)');
title('Spectral Density with Windowing');
```

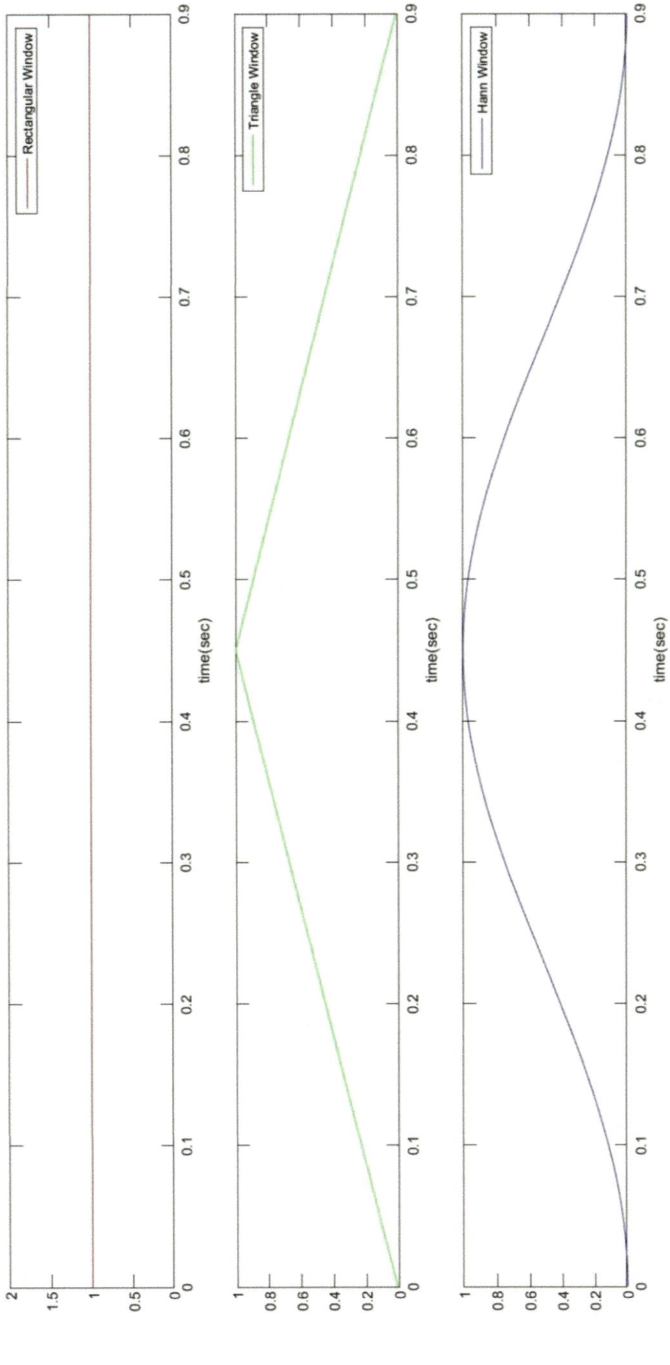

FIGURE 5.18 Various windowing functions.

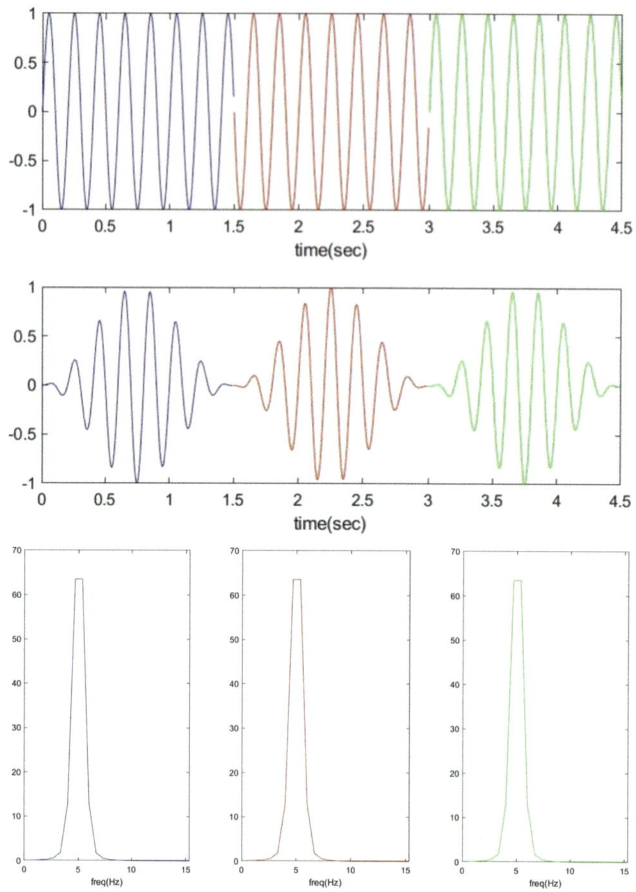

FIGURE 5.19 Application of windowing and reduction of the effect of the two end values of the block, calculation of Fourier transform, and its averaging.

5.4.4.1 Empirical Estimation of Natural Frequencies and Mode Shapes

Empirically estimating the natural frequencies and mode shapes of a system is crucial in fields such as structural health monitoring and vibration analysis. One effective method for obtaining these estimates is by analysing the cross-spectral density (CPSD) function of the free vibration segment of measured signals, using a reference signal. This approach helps identify the dominant frequencies of the system and their corresponding mode shapes, providing valuable insight into its dynamic behaviour.

The method relies on calculating the cross-spectral density (CPSD) between two signals: the measured signal X and the reference signal Y. The CPSD reveals the degree of correlation between these signals in the frequency domain. This correlation is especially helpful in identifying resonance frequencies and the dynamic modes associated with these frequencies.

In practical applications, such as when using MATLAB for signal analysis, the CPSD between two vectors X and Y can be computed using the function cpsd. This function requires several parameters to be defined for proper frequency analysis:

- Sampling Rate (F_s): The rate at which the signals were sampled.
- Block Length (NFFT): The length of the segment used for each Fourier transform computation.
- Overlap (NOVERLAP): The number of overlapping samples between consecutive blocks.
- Window Function (WINDOW): A windowing function (e.g., Hann, Hamming, or Blackman) applied to the segment to minimize frequency leakage and improve spectral estimation.

The MATLAB syntax for computing the CPSD is as follows:

```
[Pxy,F] = cpsd(X, Y, WINDOW, NOVERLAP, NFFT, Fs)
```

In this code:

- Pxy represents the complex CPSD vector, which contains the frequency-domain information of the correlation between the signals X and Y.
- F is the corresponding frequency vector, containing the frequency values at which the cross-spectral density is calculated.

The output of the cpsd function provides the frequency-dependent correlation, helping to identify dominant modes and resonant frequencies of the system. This is crucial in empirical modal analysis, where the goal is to characterize the natural frequencies and mode shapes based on measured data.

By analysing the cross-spectral density function, one can not only determine the system's resonant frequencies but also gain insight into the mode shapes associated with these frequencies. This is particularly useful in structural vibration analysis, where identifying mode shapes is essential for diagnosing structural integrity and predicting response behaviour under different loading conditions.

For example for Neka bridge, the cross-spectral density for free vibration sections of various channels (5, 8, 14, 20, and 23) relative to the reference channel 12 is computed. These channels correspond to different positions along the bridge's structure, with distances from the left-side bearing:

- Channel 5: 6 m in the southern truss
- Channel 8: 16 m in the southern truss
- Channel 12: 18 m in the southern truss (reference)
- Channel 14: 18 m in the northern truss
- Channel 20: 26 m in the southern truss
- Channel 23: 30 m in the southern truss

In MATLAB, the cross-spectral density is calculated as follows:

```
[pf05, fr] = cpsd(CH12(16335:end), CH05(16335:end), [], [], [], 2000);
[pf08, fr] = cpsd(CH12(16335:end), CH08(16335:end), [], [], [], 2000);
[pf14, fr] = cpsd(CH12(16335:end), CH14(16335:end), [], [], [], 2000);
[pf20, fr] = cpsd(CH12(16335:end), CH20(16335:end), [], [], [], 2000);
[pf23, fr] = cpsd(CH12(16335:end), CH23(16335:end), [], [], [], 2000);
```

The magnitude of the cross-spectral density for the first 100 frequency bins can be plotted to visualize the spectral content:

```
plot(fr(1:100), abs(pf05(1:100)), 'r-', fr(1:100), abs(pf08(1:100)), 'b-', ...
    fr(1:100), abs(pf12(1:100)), 'y-', fr(1:100), abs(pf14(1:100)), 'c-', ...
    fr(1:100), abs(pf20(1:100)), 'k-', fr(1:100), abs(pf23(1:100)), 'g-')
legend('CH05', 'CH08', 'CH12', 'CH14', 'CH20', 'CH23')
```

This will generate a plot similar to Figure 5.20, which shows the magnitude of the cross-spectral density for the free vibration sections of the measured signals relative

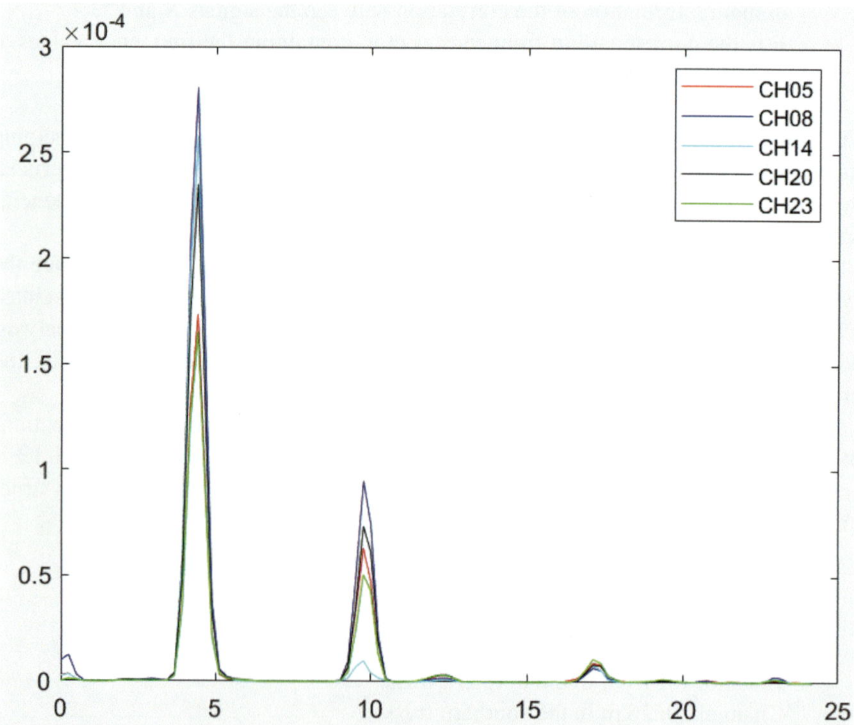

FIGURE 5.20 Magnitude of the cross-spectral density function of the free vibration section of vertical acceleration for five measurement channels relative to reference channel number 12.

to channel 12 (the reference channel). Compared to Figure 5.15, considering that the cross-correlation function has been calculated and averaging has been performed, Figure 5.20 exhibits more distinct peaks. The frequencies of 4.3, 9.3, 12.2, and 17.6 Hz can be identified in the figure.

To empirically estimate the sign of the mode shapes, the phase information of the CPSD function can be leveraged. By analysing the phase differences between the measured signals and a reference signal, we can infer whether the mode shapes are real (standing wave) or complex (moving wave) at specific frequencies. This method is particularly useful for distinguishing between different types of vibration modes in structural analysis.

A real mode shape corresponds to a standing wave, where the displacement shape at each point on the structure remains fixed with respect to the mode's frequency. For such modes, the phase difference between the signals at the natural frequency will either be 0 or 180 degrees. This means that the signals in different locations of the structure are perfectly in-phase or anti-phase, indicating a standing wave behaviour.

The phase difference of 0 degrees implies that the signals are in perfect synchronization, while a phase difference of 180 degrees means that they are exactly opposite in sign. Both of these scenarios are characteristic of a real mode shape, where the deformation does not propagate along the structure.

In contrast, a complex mode shape corresponds to a moving wave, where the displacement of the structure varies continuously over time and space. For a complex mode shape, the phase difference between signals will not be exactly 0 or 180 degrees. Instead, the phase difference will vary, indicating that the displacement is changing dynamically as the wave propagates through the structure.

The variation in the phase difference suggests that the signals are not entirely synchronized, nor are they perfectly anti-phase. Instead, the phase difference fluctuates, signifying that the deformation is propagating across the structure and is characteristic of a moving wave mode shape.

The phase difference between signals is essential for distinguishing between real and complex mode shapes. By examining the phase at the natural frequencies, one can determine the type of mode shape.

If the phase difference at a particular frequency is close to 0 or 180 degrees, it suggests a standing wave, corresponding to a real mode shape.

- If the phase difference shows variation and is not confined to 0 or 180 degrees, it indicates a moving wave, corresponding to a complex mode shape.

Continuing on Neka bridge's example, the phase for different channels can be plotted in the frequency range 3–6 Hz to examine the phase behaviour. In this case, the peak occurs at 4.4 Hz:

```
plot(fr(1:100), angle(pf05(1:100)), 'r-', fr(1:100), angle(pf08(1:100)), 'b-',
    fr(1:100), angle(pf14(1:100)), 'c-', fr(1:100), angle(pf20(1:100)), 'k-',
    fr(1:100), angle(pf23(1:100)), 'g-')
legend('CH05', 'CH08', 'CH14', 'CH20', 'CH23')
xlim([11 13.5])
```

This code creates plots that are shown in Figure 5.21. This plot shows the phase behaviour of the cross-spectral density function for the various channels with respect to the reference channel (Channel 12). The phase data will help in determining whether the mode shape is real or complex.

At the frequency of 4.4 Hz (first vertical mode shape of the bridge), which corresponds to the 19th index of the cross-spectral density function, both the magnitude and phase data can be used to plot the mode shape. The magnitude is extracted from the left plot, and the sign is determined from the phase curve on the right.

The following code illustrates the magnitude and phase for constructing the mode shape, which will generate Figure 5.22, illustrating the mode shape at 4.4 Hz. The mode shape is plotted in such a way that, considering channels 5, 8, 12, 20, and

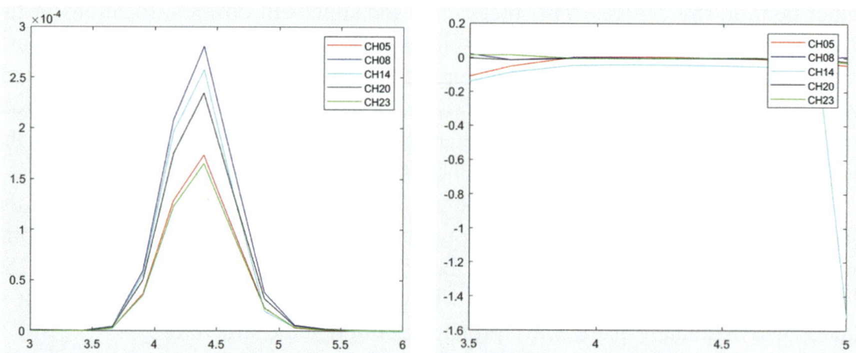

FIGURE 5.21 Magnitude and phase of the cross-spectral density function of the free vibration section of vertical acceleration for five measurement channels relative to reference channel number 12 in the range of 3–6 Hz (peak at 4.4 Hz).

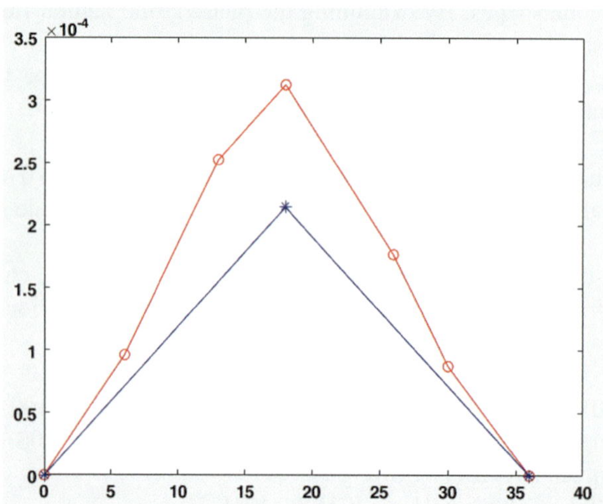

FIGURE 5.22 Mode shape at 4.4 Hz.

23 measure at 6, 13, 18, 26, and 30 m from the support on the southern truss, and channel 14 is installed at 18 m (mid-span) on the northern truss, and given that the measurements are in phase and therefore have the same sign, the cross-spectrum magnitude at the examined frequency is plotted at the accelerometer installation positions.

```
plot([0 6 13 18 26 30 36], abs([0 ppf05(19) ppf08(19) ppf12(19) ppf20(19)
ppf23(19) 0]), 'r-o', ...
    [0 18 36], [0 ppf14(19) 0], 'b-*')
```

Similarly, at 12.4 Hz (second vertical mode shape of the bridge), where the peak occurs at the 52nd index of the cross-spectral density function, both magnitude and phase data can be used to plot the mode shape which is shown in Figures 5.23 and 5.24.

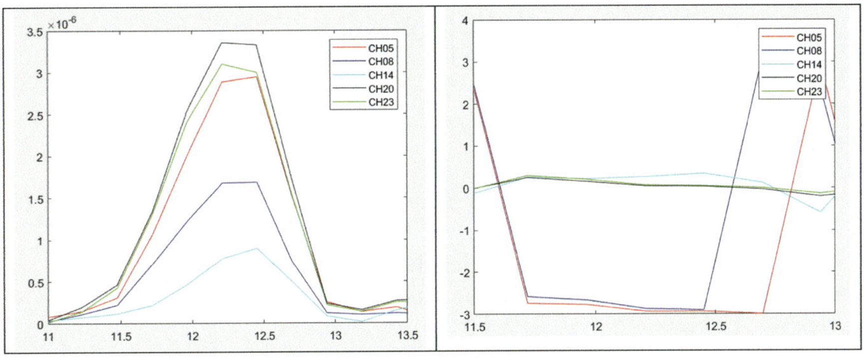

FIGURE 5.23 Magnitude and phase of the cross-spectral density function of the free vibration section of vertical acceleration for five measurement channels relative to reference channel number 12 in the range of 11 to 13.5 Hz (peak at 12.4 Hz).

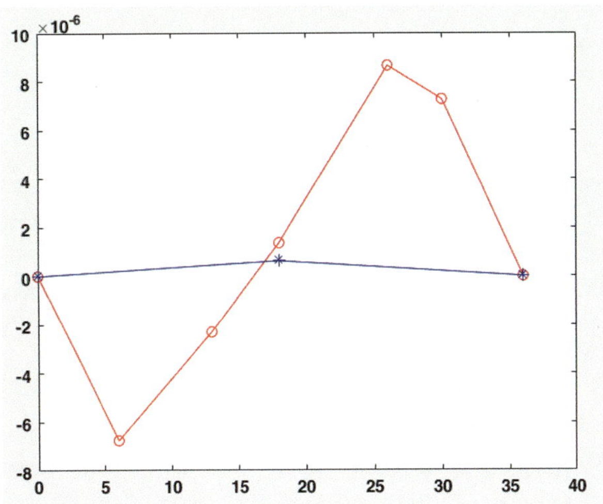

FIGURE 5.24 Mode shape at 12.4 Hz.

5.4.5 Empirical Estimation of Damping

Damping refers to a structure's capacity to dissipate energy imparted by external forces. Although the mechanisms of damping are highly diverse, their overall effects on vibration are typically summarized using an equivalent viscous damping ratio (ζ), expressed as a fraction of critical damping. For a damping ratio of 1%, the amplitude is halved after 11 cycles, while for a damping ratio of 10%, the amplitude is halved after just 1 cycle. Several methods exist for estimating damping, including the stochastic subspace identification method and the logarithmic decrement method.

In the logarithmic decrement method, the equivalent damping ratio can be estimated by measuring the amplitude reduction after n cycles, according to the following equation and as shown in Figure 5.25:

$$\ln\left(\frac{x_1}{x_{n+1}}\right) = 2\pi \times n \times \zeta$$

where

- x_1 and x_{n+1} are the vibration amplitudes at the first and (n+1)th cycle, respectively.
- n is the number of cycles over which the decay is measured.
- ζ is the equivalent viscous damping ratio.

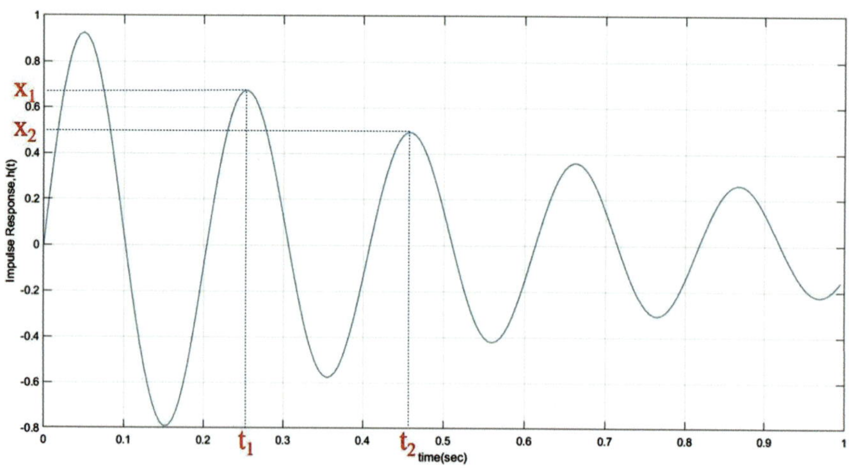

FIGURE 5.25 Damping estimation using the logarithmic decrement method $\ln\left(\frac{x_1}{x_2}\right) = 2\pi \times \zeta$.

Figure 5.26 demonstrates a SDOF system with a natural frequency of 5.2 Hz, showcasing its vibrational behaviour under different damping ratios of 1% and 5% [1]. This comparison highlights how damping influences the rate of amplitude decay and the overall energy dissipation of the system.

In a real-world application, the damping ratio can be estimated from measured signals by first isolating the vibration at the natural frequency of interest. This process typically involves passing the signal through a bandpass filter to isolate the desired frequency range. For example, using MATLAB, a fourth-order Butterworth bandpass filter can be applied as follows:

```
[b, a] = butter(4, [f_low/Fs, f_high/Fs])
filtered_signal = filter(b, a, original_signal)
```

Once filtered, the free vibration portion of the signal is analysed to measure the amplitude reduction over several cycles. In Figure 5.27 the damping ratio for vertical acceleration at the centre span of the northern truss is estimated in the 2–6 Hz frequency range using the logarithmic decrement method. The fourth-order Butterworth

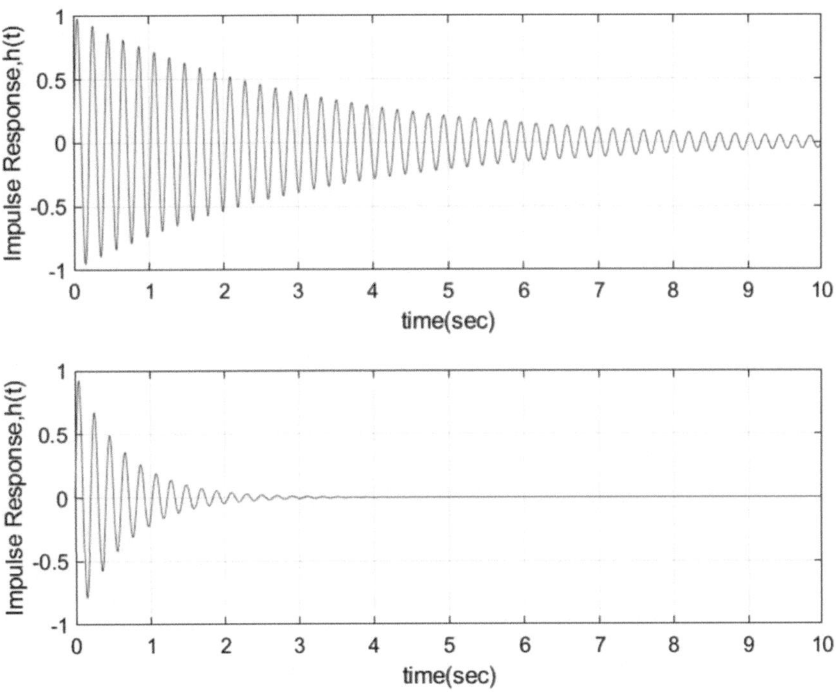

FIGURE 5.26 Single degree-of-freedom system with a natural frequency of 5.2 Hz and damping ratios of 1% and 5%.

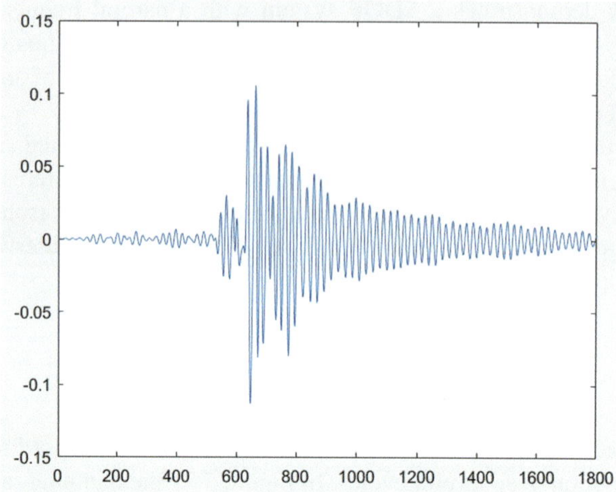

FIGURE 5.27 Damping estimation from Channel 14 (vertical acceleration at the centre span of the northern truss) in 2–6 Hz range.

bandpass filter isolates the natural frequency range of a measurement with a sampling frequency of 100 Hz:

```
[b, a] = butter(4, [2/50, 6/50])
CH14rf = filter(b, a, CH14r)
```

Here, the amplitude decreases from 0.04521 to 0.0201 over 10 cycles, leading to an equivalent damping ratio of 1.3% at a frequency of 4.4 Hz:

$$\zeta = \frac{\ln\left(\dfrac{0.04521}{0.0201}\right)}{2\pi \times 10} = 1.3\%$$

Similarly, Figure 5.28 estimates the damping ratio for the 11–13.5 Hz range. The signal is filtered as follows:

```
[b, a] = butter(4, [11/50, 13.5/50])
CH14rf = filter(b, a, CH14r)
```

Over 10 cycles, the amplitude decreases from 0.01913 to 0.00371, yielding a damping ratio of 2.6% at a frequency of 12.4 Hz.

These examples illustrate how the logarithmic decrement method can estimate the damping ratio by analysing amplitude reduction. By systematically applying this approach, damping ratios can be determined for all the natural frequencies of interest, enabling engineers to characterize the energy dissipation behaviour of a structure comprehensively.

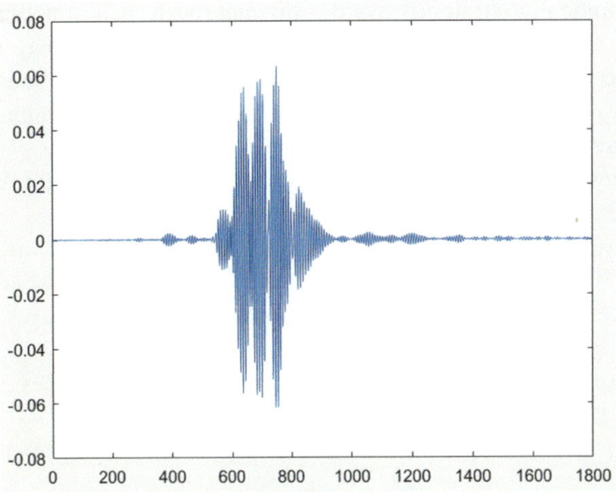

FIGURE 5.28 Damping estimation from Channel 14 (vertical acceleration at the centre span of the northern truss) in 11–13.5 Hz range.

5.5 CONVERSION OF ACCELERATION TO DISPLACEMENT

From a theoretical standpoint, displacement can be derived by integrating velocity, while velocity can be obtained by integrating acceleration. These relationships are fundamental in vibration analysis, as they provide insight into the dynamic behaviour of a system based on measured quantities. However, translating these theoretical concepts into practice comes with challenges, particularly related to the accuracy of integration.

The precision of integration is highly dependent on the sensitivity and quality of the accelerometer or velocity sensor used for data acquisition. This is especially critical at frequencies near zero (low-frequency range), where sensor noise and drift significantly impact the reliability of the integrated signal. To mitigate these inaccuracies, practical integration is typically constrained to frequencies above a certain threshold. This threshold is determined by the characteristics of the vibration sensor employed, such as its frequency response, noise floor, and resolution [1].

Since the first derivative of $\sin(\omega t)$ is $\omega\cos(\omega t)$, and its second derivative is $-\omega^2\sin(\omega t)$, integration in the frequency domain can be carried out by dividing the frequency content by $-\omega^2$ at each frequency ω.

In the frequency domain, the process of integration involves specific mathematical transformations. Since the derivative of $\sin(\omega t)$ is $\omega\cos(\omega t)$, and the second derivative is $-\omega^2\sin(\omega t)$, integration can be performed by reversing these operations. For instance, integrating acceleration to obtain displacement requires dividing the frequency content of the signal by $-\omega^2$ at each angular frequency ω. Similarly, integrating velocity to derive displacement involves dividing the frequency content by ω. These operations effectively adjust the amplitude of the signal in the frequency domain, ensuring consistency with the physical relationships between displacement, velocity, and acceleration.

While theoretically straightforward, this approach has practical limitations. Dividing by $-\omega^2$ amplifies low-frequency components, which may include sensor noise or artefacts introduced during measurement. Consequently, the integration process is typically coupled with signal filtering techniques, such as high-pass filters, to suppress low-frequency noise and improve accuracy.

The following MATLAB function converts an acceleration vector A to a displacement vector E2:

```
% MATLAB Function to Convert Acceleration to Displacement
function E2 = AccelerationToDisplacement(A, Fs, freq_cutoff)
  % Inputs:
  % A - Acceleration vector
  % Fs - Sampling frequency
   % freq_cutoff - Minimum frequency for integration

  N = length(A);
  f = (0:N-1)*(Fs/N); % Frequency vector
  omega = 2*pi*f;     % Angular frequency

  % Frequency domain transformation
  A_fft = fft(A);
  Integrator = zeros(size(A_fft));

  % Perform integration above cutoff frequency
  for k = 1:N
     if f(k) >= freq_cutoff
        Integrator(k) = -1/(omega(k)^2); % Integration factor
     end
  end

  % Apply integration filter
  Displacement_fft = A_fft .* Integrator;

  % Transform back to time domain
  E2 = real(ifft(Displacement_fft));
end
```

For instance, in the dynamic testing of the Neka bridge, integration from acceleration to displacement was accurately performed for frequencies above 1.59 Hz. Integration can be executed in the time domain, frequency domain, or using integrator filters. This method was applied to integrate the acceleration of the mid-span of the southern truss for frequencies above 1.59 Hz to compute displacement using the following MATLAB function. The results were compared to displacement measurements obtained directly, which are shown in Figure 5.29.

```
% Main script to process acceleration data and convert to displacement

% Inputs
Fs = 2000; % Sampling frequency in Hz
freq_cutoff = 1.59; % Minimum frequency for integration in Hz
upper_freq_limit = 30; % Upper frequency limit in Hz

% Load or define the acceleration data matrix A
% A(:,1:27) contains acceleration values in g (gravitational units)
A(:,1:27) = 9.81 * 1000 * A(:,1:27); % Convert to mm/s²

[n1, n2] = size(A); % Get the size of the input data
E = zeros(size(A)); % Initialize the displacement matrix
E2 = zeros(size(A)); % Initialize filtered displacement matrix

for col = 1:n2
    % Process each column (channel) separately
    channel_data = A(:, col);

    % Convert acceleration to displacement (full frequency range)
    displacement_full = AccelerationToDisplacement(channel_
 data, Fs, 0);

    % Convert acceleration to displacement (filtered frequency range)
    displacement_filtered = AccelerationToDisplacement(channel_
 data, Fs, freq_cutoff);

    % Remove frequencies above the upper limit
    freq_vector = (0:n1-1) * (Fs / n1); % Frequency vector
    fft_data = fft(displacement_filtered);
    fft_data(freq_vector > upper_freq_limit) = 0; % Zero out
 frequencies above the limit
    displacement_filtered = real(ifft(fft_data));

    % Remove DC offset from displacement signals
    displacement_full = displacement_full - mean
 (displacement_full(end-Fs:end));
    displacement_filtered = displacement_filtered - mean
 (displacement_filtered(end-Fs:end));

    % Store results
    E(:, col) = displacement_full; % Full range displacement
    E2(:, col) = displacement_filtered; % Filtered displacement
end

% At this point, E contains displacement from the entire fre-
quency range
% E2 contains displacement from the filtered frequency range
(1.59 Hz to 30 Hz)
```

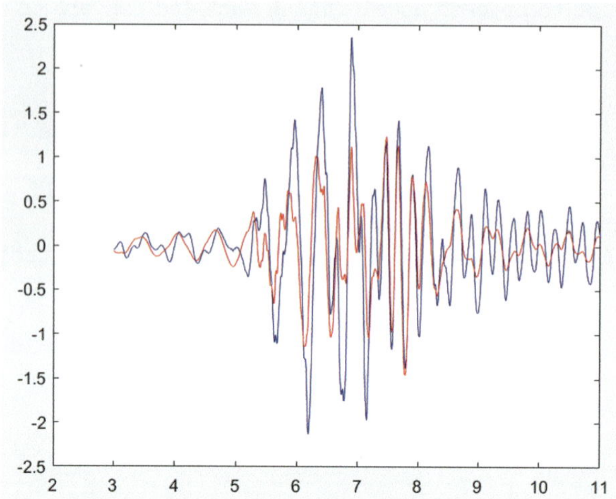

FIGURE 5.29 Comparison of measured vertical displacement from sensor channel 35 (red signal) and displacement derived from double integration of channel 12 (blue signal) during the passage of a diesel locomotive over the Neka Bridge within the frequency range of 1.59–30 Hz.

Figure 5.29 shows a comparison between the vertical displacement data directly measured from Channel 35 (vertical displacement at the mid-span of the southern truss) and the displacement data derived from double integration of the vertical acceleration signal in Channel 12 (vertical acceleration at the mid-span of the southern truss). The comparison focuses on the frequency range from 1.59 to 30 Hz, which was selected to filter out low-frequency noise and capture the relevant dynamic response of the bridge. This figure illustrates the close correlation between the derived and measured displacements, confirming the accuracy of the integration process in modelling the bridge's response to dynamic loading from the passing locomotive.

5.6 FREQUENCY RESPONSE FUNCTION

The frequency response function (FRF) is a fundamental concept in structural dynamics and vibration analysis. It describes how a system responds to an external force at different frequencies, and is crucial for understanding the behaviour of structures under dynamic loading conditions, such as vibrations caused by machinery, traffic, wind, or earthquakes.

The FRF is a measure of the system's output (displacement, velocity, or acceleration) relative to the input (force) at each frequency. It provides insights into the dynamic behaviour of the structure, such as identifying natural frequencies where resonance might occur. FRFs are typically used to analyse how a structure or component vibrates when subjected to periodic forces.

In essence, the FRF is a ratio of the amplitude of the output (e.g., displacement) to the amplitude of the input (force), expressed as a function of frequency, which can be represented mathematically as:

$$H(\omega) = \frac{X(\omega)}{F(\omega)}$$

where

- $H(\omega)$ is the frequency response function at angular frequency ω,
- $X(\omega)$ is the output displacement (or velocity or acceleration) at frequency ω,
- $F(\omega)$ is the input force at frequency ω.

To understand the FRF, we start with the concept of a single-degree-of-freedom (SDOF) system (Figure 5.30). This is the simplest model of a vibrating system, where the structure is assumed to move in only one direction (e.g., along the length of a beam or a column) and has one natural mode of vibration.

For an SDOF system, the response to an external harmonic force can be described by the following equation of motion:

$$M\ddot{x} + C\dot{x} + Kx = f(t)$$

Here:

- M is the mass of the structure (kg),
- C is the damping coefficient (Ns/m),
- K is the stiffness of the structure (N/m),
- $x(t)$ is the displacement as a function of time (m),
- $f(t)$ is the applied force as a function of time (N).

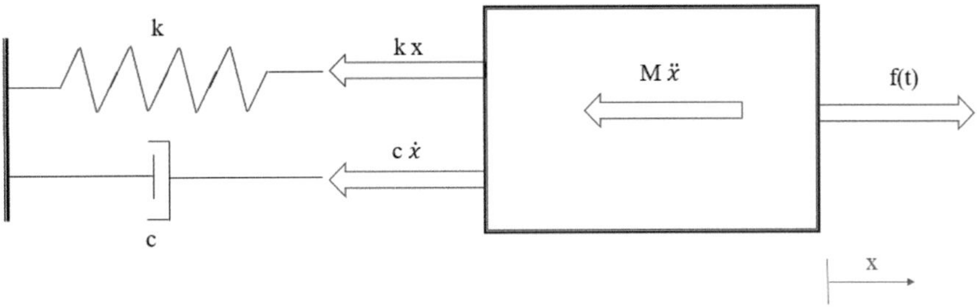

FIGURE 5.30 Free body diagram of a single-degree-of-freedom system.

This differential equation describes the dynamic response of a system to an applied force, considering mass, damping, and stiffness. When analysing dynamic behaviour, we are usually interested in the system's response at various frequencies, so this equation is typically transformed into the frequency domain using the Fourier transform or Laplace transform as described before.

The transfer function, or frequency response function, of a system relates the input force $f(t)$ and the resulting displacement $x(t)$ at different frequencies. To derive this, we assume that the input force is harmonic, that is, it oscillates at a specific frequency ω. We express the displacement and the force in terms of complex exponentials, $e^{i\omega t}$, where i is the imaginary unit and ω is the angular frequency.

In the frequency domain, the transfer function becomes:

$$H(\omega)=\frac{X(\omega)}{F(\omega)}=\frac{1}{-M\omega^2+iC\omega+K}$$

in which ω is the frequency of the applied force, and the rest of the parameters are as defined before. This equation shows how the system's displacement $X(\omega)$ is affected by the frequency of the input force. The term $iC\omega$ represents the effect of damping, and $-M\omega^2+K$ represents the stiffness and inertia of the system.

As explained before, the natural frequency ω_n and the damping ratio ζ of the system are as follows:

$$\omega_n=\sqrt{\frac{K}{M}}$$

$$\zeta=\frac{C}{2\sqrt{K.M}}$$

where C is the damping coefficient, and the rest of the parameters are as described before. Thus, the frequency response function at ω (the frequency of the applied force) can be expanded as follows:

$$H(\omega)=\frac{X(\omega)}{F(\omega)}=\frac{\frac{1}{M}}{-\omega^2+i2\zeta\omega_n\omega+\omega_n^2}=\frac{\frac{1}{K}}{-(\frac{\omega}{\omega_n})^2+i2\zeta\frac{\omega}{\omega_n}+1}$$

This form shows that the system's response changes as the frequency of the applied force varies. At resonance, where the driving frequency is equal to the natural frequency of the system, the displacement can become large, potentially causing structural damage. Figure 5.31 presents the FRF of a SDOF system for different damping ratios.

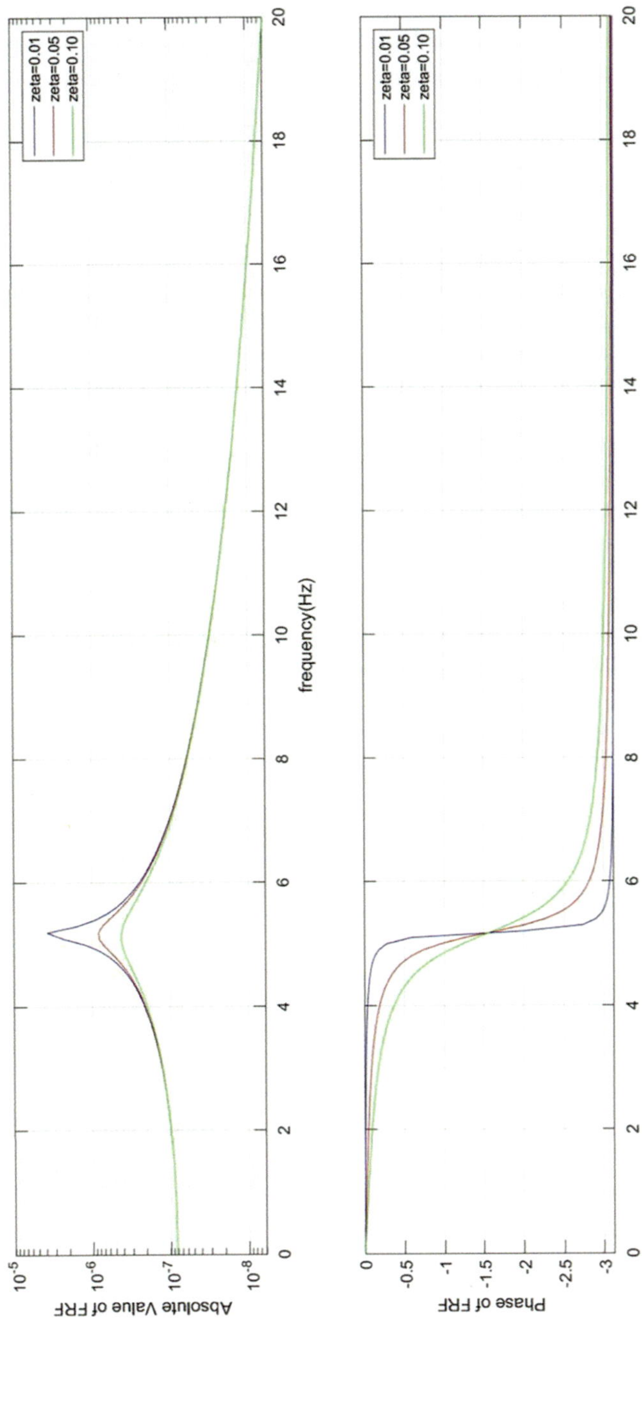

FIGURE 5.31 Frequency response function of a single-degree-of-freedom system with different damping ratios.

In a multi-degree-of-freedom (MDOF) system, the dynamics are more complex than those of an SDOF system due to the interaction between multiple masses and stiffness elements. The system's behaviour can be described by a set of coupled equations that must be solved simultaneously. These equations can be expressed in matrix form without damping as follows:

$$\left[-[M]\omega^2 + [K]\right]\{X(\omega)\} = \{F(\omega)\}$$

where

- M is the mass matrix (which defines the inertial properties of the system),
- K is the stiffness matrix (which defines the stiffness of the system),
- $X(\omega)$ is the vector of displacements (describing the system's response),
- $F(\omega)$ is the force vector (describing the external forces applied to the system).

The equation expresses how the forces applied to the system lead to a displacement response, taking into account the mass and stiffness of the system. The left-hand side represents the system's inertial and elastic behaviour, while the right-hand side represents the external forces. To solve for the displacement $X(\omega)$, we must compute the inverse of the matrix $-[M]\omega^2 + [K]$.

Modal analysis is a technique used to simplify the dynamics of a MDOF system by decomposing the system into its fundamental modes of vibration. Each mode corresponds to a specific frequency and shape of vibration. In the absence of damping, these modes are orthogonal with respect to the mass and stiffness matrices, which simplifies the solution of the system. For the system to be considered "orthogonal," the following relationships must hold for the modal shape vectors ψ:

$$[\Psi]^T[\mathrm{M}][\Psi] = \begin{bmatrix} m_1 & \cdots & 0 \\ \vdots & \ddots & \vdots \\ 0 & \cdots & m_N \end{bmatrix}$$

$$[\Psi]^T[K][\Psi] = \begin{bmatrix} k_1 & \cdots & 0 \\ \vdots & \ddots & \vdots \\ 0 & \cdots & k_N \end{bmatrix}$$

where:

- Ψ is the modal matrix (containing the modal shape vectors),
- $m_1, m_2, \ldots, m_n$ are the modal masses,
- $k_1, k_2, \ldots, k_n$ are the modal stiffnesses.

This orthogonality allows the modal shapes to be decoupled from one another, and each mode can be analysed independently, which simplifies the problem. The diagonal form of these equations means that each mode contributes separately to the system's overall dynamic behaviour.

From the previous relationships, we can calculate the natural frequencies of the system, which are fundamental to understanding the system's vibration behaviour. The natural frequencies ω_1, ω_2, ..., ω_n are related to the modal stiffness and modal mass matrices by the equation:

$$\omega_r^2 = k_r / m_r$$

This shows how the natural frequencies ω_r^2 are determined by the system's modal mass and stiffness. These frequencies correspond to the system's resonant frequencies, at which the system will vibrate most easily when excited.

To obtain a transfer function for a specific mode, we can use the modal decomposition approach, which involves projecting the system's behaviour onto the modal shapes Ψ. Finally, the resulting transfer function of the system, accounting for both stiffness, mass, and damping, can be expressed as:

FRF matrix of MDOF system ($H(\omega)$) can be calculated as shown below:

$$\{X\} = \left([K] - \omega^2 [M]\right)^{-1} \{F\}$$

$$\{X\} = [H(\omega)]\{F\}$$

$$\left([K] - \omega^2 [M]\right) = [H(\omega)]^{-1}$$

$$[\psi]^{\mathrm{T}} \left([K] - \omega^2 [M]\right)[\psi] = [\psi]^T [H(\omega)]^{-1} [\psi]$$

$$\left[\left(k_r - \omega^2 m_r\right)\right] = [\psi]^{\mathrm{T}} [H(\omega)]^{-1} [\psi]$$

$$[H(\omega)] = [\psi]^T \left[\left(k_r - \omega^2 m_r\right)\right]^{-1} [\psi]$$

$$[H(\omega)] = [\psi] \begin{bmatrix} k_1 - m_1\omega^2 & \cdots & 0 \\ \vdots & \ddots & \vdots \\ 0 & \cdots & k_N - m_N\omega^2 \end{bmatrix}^{-1} [\psi]^{\mathrm{T}}$$

$$H_{jk}(\omega) = \sum_{r=1}^{N} \frac{\psi_{jr} \cdot \psi_{kr}}{k_r - m_r\omega^2}$$

where:

- X_j is the displacement response at the j-th degree of freedom,
- F_k is the applied force at the k-th degree of freedom,
- ψ_{jr} and ψ_{kr} are the components of the modal shape vector ψ,
- k_r and m_r are the stiffness and mass of the r-th mode,
- ω is the angular frequency.

This transfer function indicates how each mode contributes to the overall system response, considering both the stiffness and mass for each mode.

When damping is present in the system, it affects the vibrational response by dissipating energy. If damping is assumed to be proportional (i.e., it can be modelled as a linear combination of the mass and stiffness matrices), then the damping matrix C can be written as:

$$[C] = \beta[K] + \gamma[M]$$

where β and γ are constants that define the proportionality between damping, stiffness, and mass. This assumption is commonly used in engineering because it simplifies the modelling of damping, especially when the exact distribution of damping in the system is unknown. The damping matrix affects the frequency response by shifting the natural frequencies and introducing energy dissipation.

As we want to have proportional damping, that is, we can decompose the equation of motion in a modal coordinate.

Similar to mass and stiffness matrices, the proportional damping matrix in modal coordinates can be expressed as:

$$[\Psi]^{\mathrm{T}}[C][\Psi] = \beta\begin{bmatrix} k_1 & \cdots & 0 \\ \vdots & \ddots & \vdots \\ 0 & \cdots & k_N \end{bmatrix} + \gamma\begin{bmatrix} m_1 & \cdots & 0 \\ \vdots & \ddots & \vdots \\ 0 & \cdots & m_N \end{bmatrix} = \begin{bmatrix} \beta k_1 + \gamma m_1 & \cdots & 0 \\ \vdots & \ddots & \vdots \\ 0 & \cdots & \beta k_N + \gamma m_N \end{bmatrix}$$

To make it easier to understand, let's consider the 2 DOF system of Neka bridge presented earlier, and calculate H_{11}. The mass, stiffness, eigenvectors and eigenvalues were calculated before, and are reproduced below:

$$M = \begin{bmatrix} 11340 & 0 \\ 0 & 11340 \end{bmatrix} \mathrm{kg},$$

$$K = \begin{bmatrix} 2.05 & 0.19 \\ 0.19 & 0.12 \end{bmatrix} \times 10^8 \, \frac{\mathrm{N}}{\mathrm{m}}.$$

$$[\psi]=\begin{bmatrix}0.93 & -0.09\\ 0.09 & 0.93\end{bmatrix}$$

$$[\psi]^{\mathrm{T}}[M][\psi]=[m_r]=\begin{bmatrix}9900 & 0\\ 0 & 9900\end{bmatrix}$$

$$[\psi]^{\mathrm{T}}[K][\psi]=[k_r]=\begin{bmatrix}1.81 & 0\\ 0 & 0.09\end{bmatrix}\times 10^8$$

$H_{11}(\omega)$ the displacement of the first mass (X_1) to the force applied to the first mass (F_1).

$$H_{11}(\omega)=\frac{X_1}{F_1}=\sum_{r=1}^{N}\frac{\psi_{1r}\cdot\psi_{1r}}{k_r-m_r\omega^2}$$

For our case with two degrees of freedom, this simplifies to:

$$H_{11}(\omega)=\frac{\psi_{11}\cdot\psi_{11}}{k_1-m_1\omega^2}+\frac{\psi_{12}\cdot\psi_{12}}{k_2-m_2\omega^2}$$

$$H_{11}(\omega)=\frac{X_1}{F_1}=\frac{(0.93)^2}{1.18\times 10^8-9900\omega^2}+\frac{(0.09)^2}{0.09\times 10^8-9900\omega^2}$$

This expression represents the frequency response function $H_{11}(\omega)$, which can now be used to analyse the system's dynamic behaviour.

Let's now solve another example in which damping is also considered. The damping matrix represents energy dissipation in the system. Here, it is assumed to be a linear combination of the stiffness and mass matrices:

$$[C]=1.2\times 10^{-5}[K]+0.556[M]=\begin{bmatrix}8765 & 228\\ 228 & 6449\end{bmatrix}$$

$$[\Psi]^{\mathrm{T}}[\mathrm{C}][\Psi]=\begin{bmatrix}7671 & 0\\ 0 & 5611\end{bmatrix}$$

The transfer function $H_{11}(\omega)$ represents the dynamic response X_1 of the first degree of freedom due to an applied force F_1 at the same degree of freedom. Using the formula:

$$H_{11}(\omega) = \frac{X_1}{F_1} = \sum_{r=1}^{N} \frac{\psi_{1r} \cdot \psi_{1r}}{k_r - m_r\omega^2 + ic_r\omega}$$

Substituting the values for Ψ, k, m, and c :

$$H_{11}(\omega) = \frac{\psi_{11} \cdot \psi_{11}}{k_1 - m_1\omega^2 + ic_1\omega} + \frac{\psi_{12} \cdot \psi_{12}}{k_2 - m_2\omega^2 + ic_2\omega}$$
$$= \frac{(0.93)^2}{1.81\times 10^8 - 9900\omega^2 + i(7671\omega)} + \frac{(0.09)^2}{0.09\times 10^8 - 9900\omega^2 + i(5611\omega)}$$

Similarly, the transfer function $H_{12}(\omega)$ represents the dynamic response X_1 of the first degree of freedom due to an applied force F_2 at the second degree of freedom. The results are also shown in Figure 5.32.

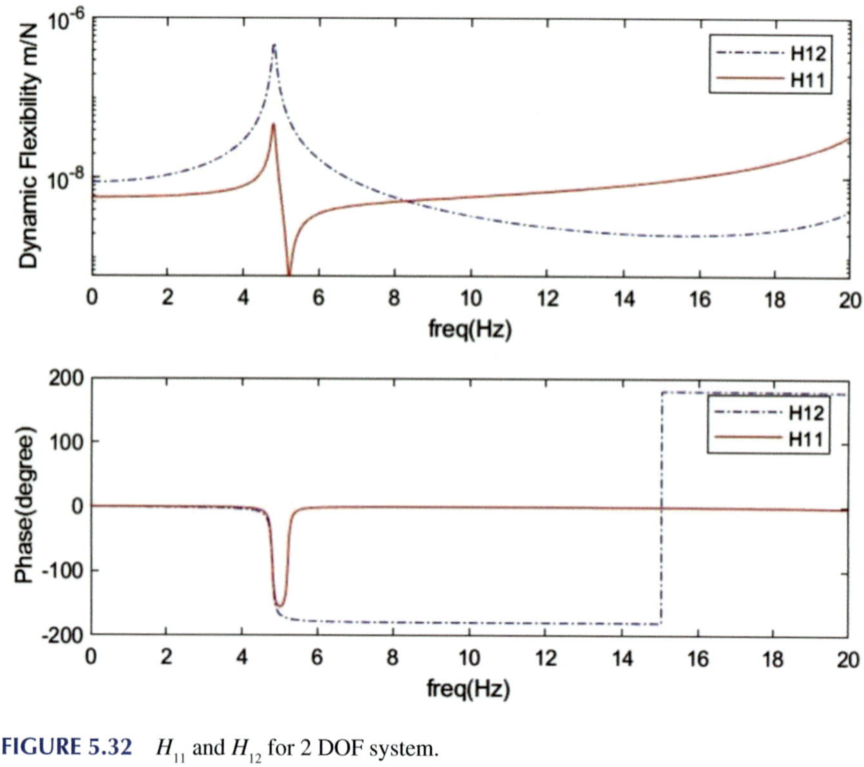

FIGURE 5.32 H_{11} and H_{12} for 2 DOF system.

$$H_{12}(\omega)=\frac{X_1}{F_2}=\sum_{r=1}^{N}\frac{\psi_{1r}\cdot\psi_{2r}}{k_r-m_r\omega^2+ic_r\omega}$$
$$=\frac{\psi_{11}\cdot\psi_{21}}{k_1-m_1\omega^2+ic_1\omega}+\frac{\psi_{12}\cdot\psi_{22}}{k_2-m_2\omega^2+ic_2\omega}$$
$$=\frac{0.93\times(-0.09)}{1.81\times10^8-9900\omega^2+i(7671\omega)}+\frac{0.09\times0.93}{0.09\times10^8-9900\omega^2+i(5611\omega)}$$

A 2D model of the Neka truss bridge was analysed to compute its natural frequencies and mode shapes in last chapter. The following frequencies were obtained:

- First natural frequency: 4.9 Hz
- Second natural frequency: 15.6 Hz
- Third natural frequency: 19.9 Hz
- Fourth natural frequency: 31.3 Hz

These natural frequencies represent the points where the structure exhibits resonance when subjected to dynamic loads. The sharp phase change (180°) observed at these frequencies in the frequency response function (FRF) is a characteristic marker of resonance. Resonance corresponds to a system pole that makes the denominator of frequency response function equal to zero, while anti-resonance corresponds to a system zero that makes the numerator of the frequency response function zero.

Using the MATLAB code provided, the FRFs $H_{24,24}$, $H_{24,22}$, and $H_{24,4}$ were computed where:

- H24,24: Force and response at DOF 24 (Node 26).
- H24,22: Force at DOF 24 (Node 26) and response at DOF 22 (Node 24).
- H24,4: Force at DOF 24 (Node 26) and response at DOF 4 (Node 6).

The FRF was computed for 50,000 frequency points using a for-loop to cover the range of interest.

```
for i = 1:50000
    a = V * inv(D - diag(i * ones(1, 47))) * V';
    a24_24(i) = a(24, 24);
    a24_22(i) = a(24, 22);
    a24_4(i) = a(24, 4);
end
```

in this code:

- V: Modal matrix of the system.
- D: Diagonal matrix of eigenvalues (natural frequencies squared).

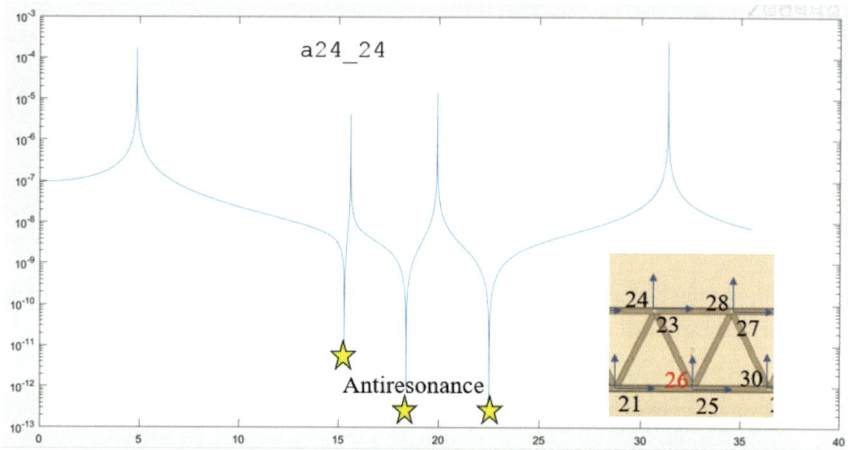

FIGURE 5.33 Magnitude of $H_{24,24}$ for the truss (degree of freedom 24, node 26).

The logarithmic plot of *H24,24* in Figure 5.33 can be created using the below MATLAB code and shows resonance peaks at the natural frequencies (4.9 Hz, 15.6 Hz, etc.).

```
semilogy(sqrt([1:50000]) / (2 * pi), abs(a24_24(1:50000)))
```

Also, the phase plot in Figure 5.34 and real part of $H_{24,24}$ in Figure 5.35 can be created using the following MATLAB code and highlights the 180° phase shift at resonance frequencies.

```
plot(sqrt([1:50000]) / (2 * pi), 180 / pi * angle(a24_
24(1:50000))), xlim([0 22])
```

In a similar manner, plots for $H_{24,22}$ and $H_{24,4}$ can be plotted, as shown in Figure 5.36 to Figure 5.39.

5.7 EMPIRICAL FREQUENCY ANALYSIS

In the previous section, we introduced the theoretical foundations of the frequency response function (FRF). This section builds on that knowledge and focuses on how to compute the FRF using real-world vibration test measurements. These tests involve applying a known input (such as a force or excitation) to a system and measuring the resulting output (such as displacement, velocity, or acceleration).

Measurements are usually in the time domain. When we talk about a system in the "time domain," we are analysing how the input and output signals change over time. The system's behaviour in the time domain is described by a convolution integral as shown below:

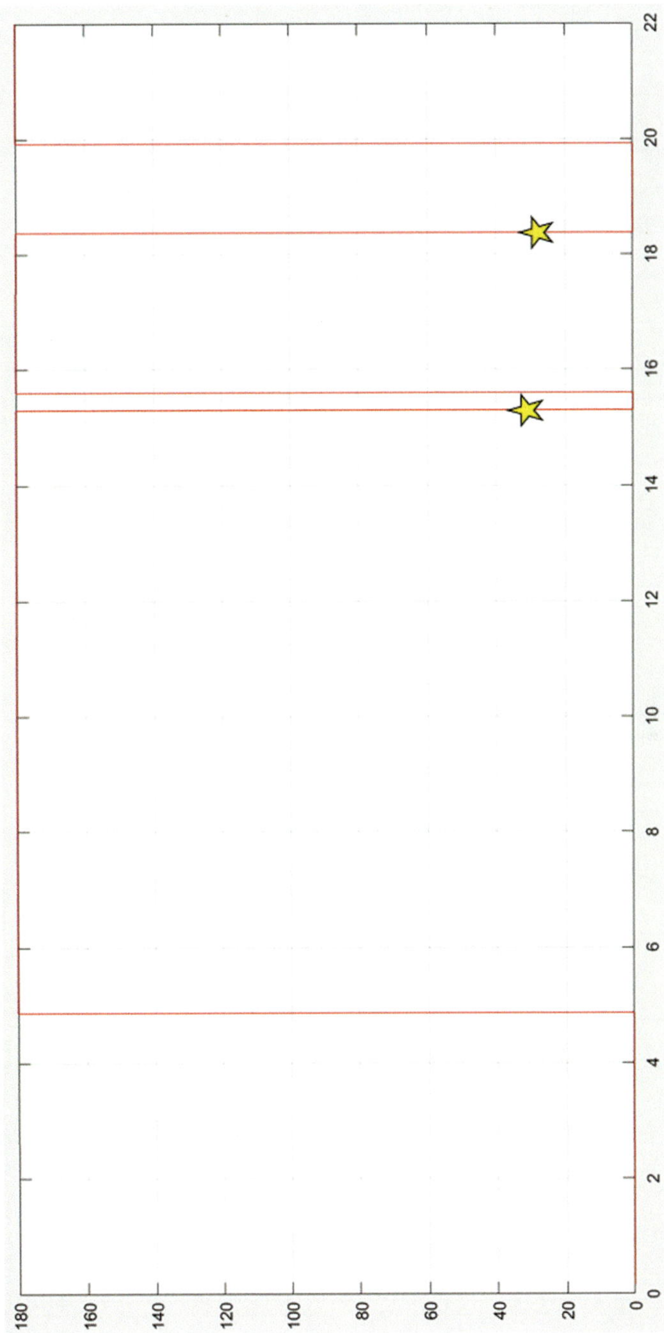

FIGURE 5.34 Phase of $H_{24,24}$ for the truss.

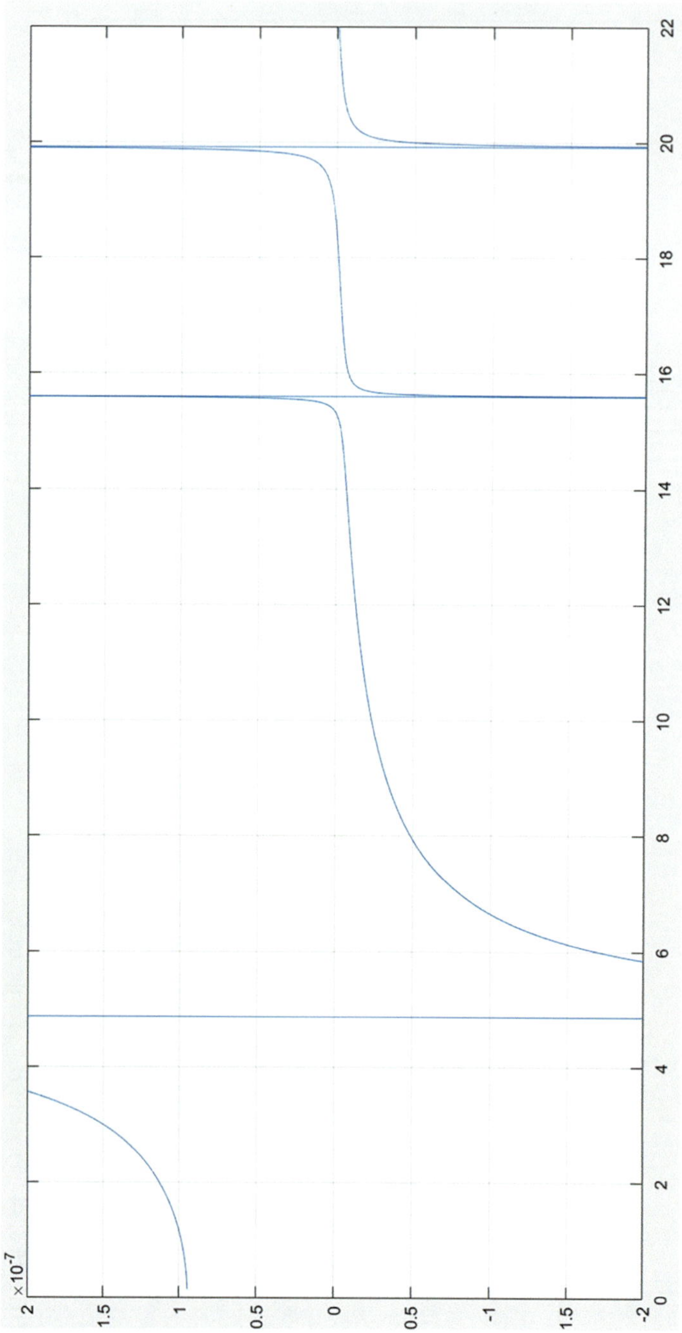

FIGURE 5.35 Real part of $H_{24,24}$ for the truss.

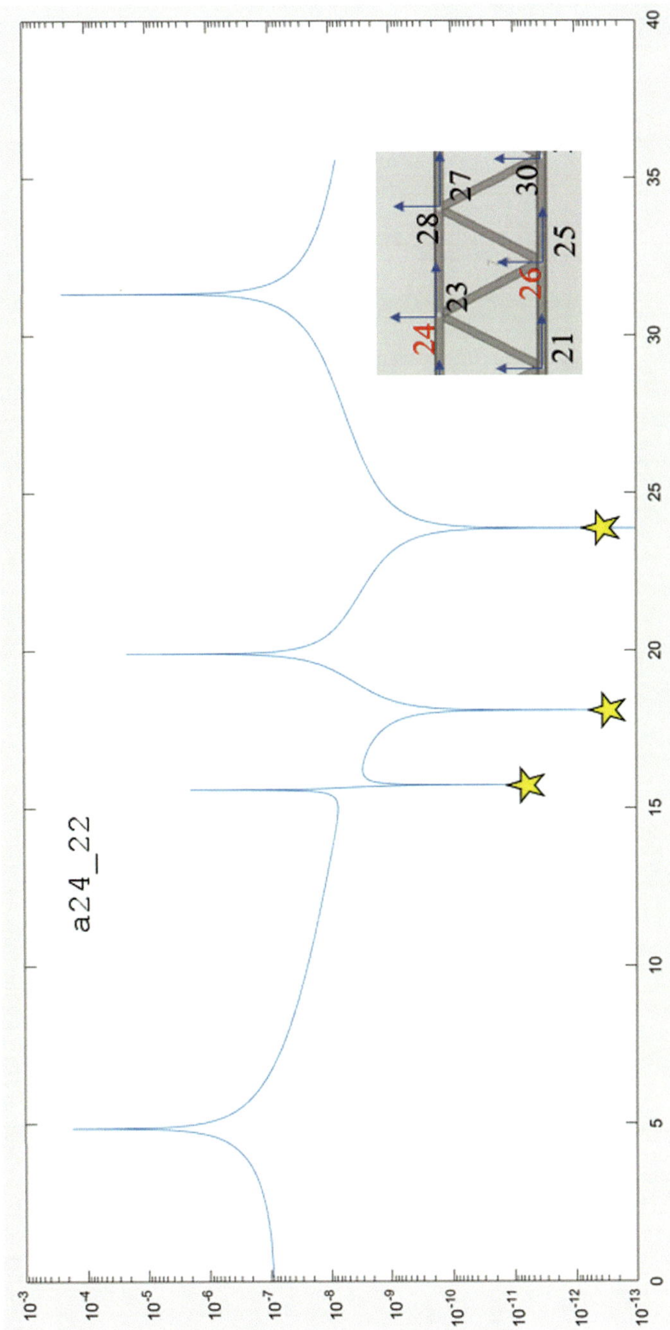

FIGURE 5.36 Magnitude of $H_{24,22}$ for the truss (force applied at degree of freedom 24, node 26; response at degree of freedom 22, node 24).

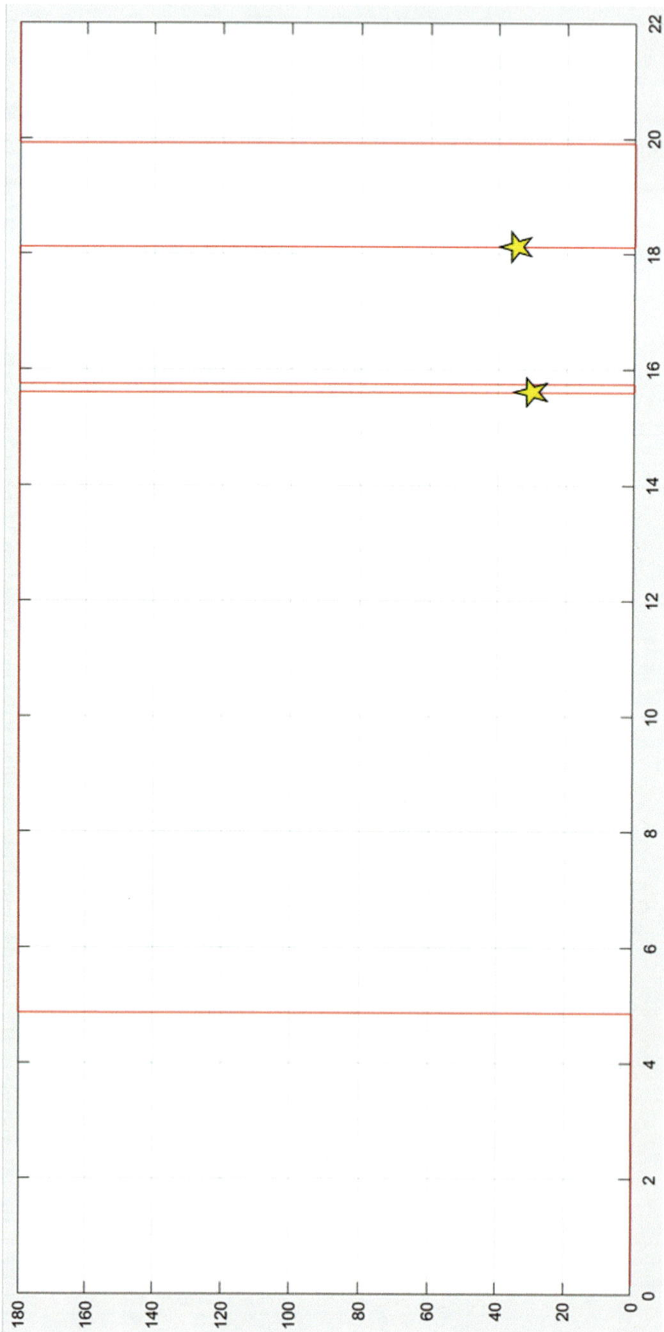

FIGURE 5.37 Phase of $H_{24,22}$ for the truss.

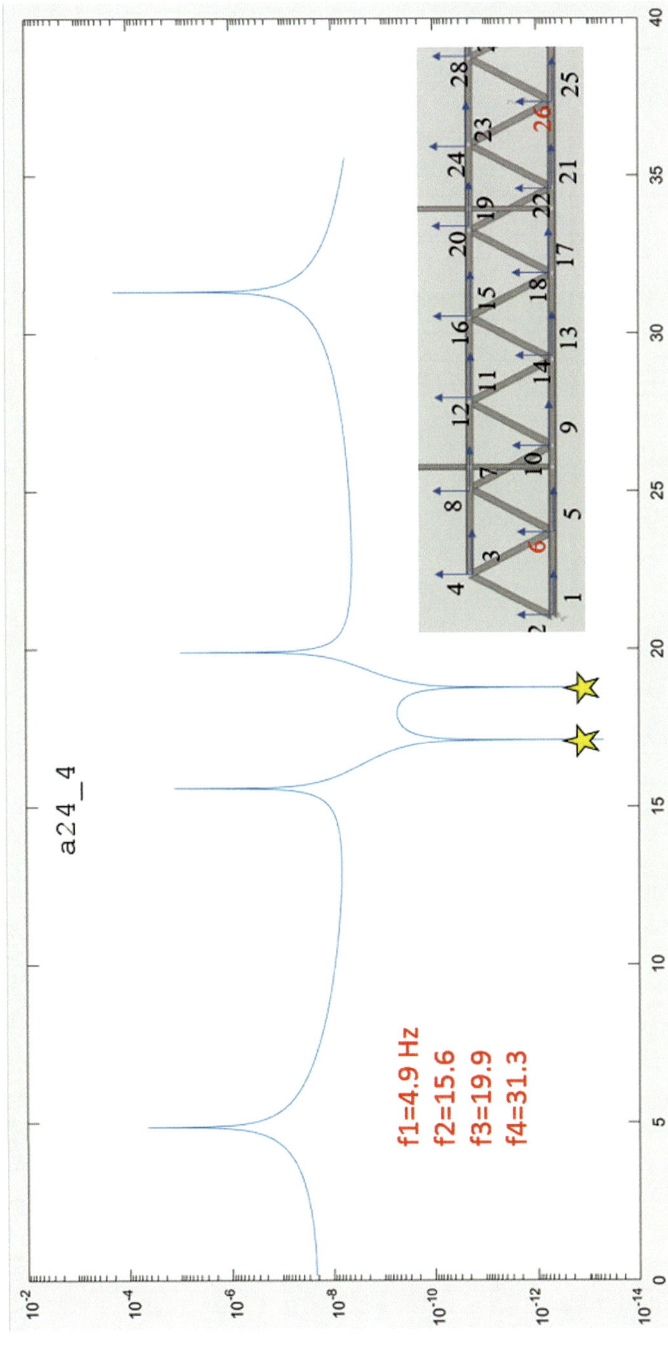

FIGURE 5.38 Magnitude of $H_{24,4}$ for the truss (force applied at degree of freedom 24, node 26; response at degree of freedom 4, node 6).

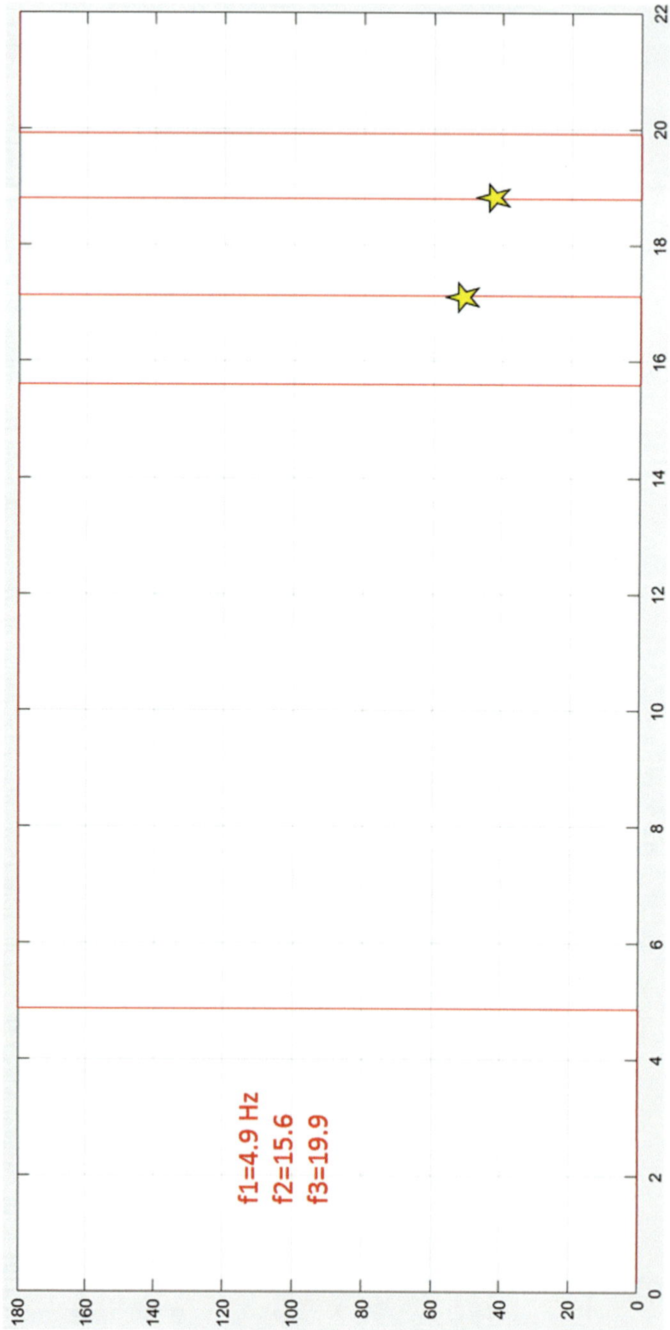

FIGURE 5.39 Phase of $H_{24,4}$ for the truss.

$$y(t) = x(t) * h(t) = \int_{-\infty}^{+\infty} x(u)h(t-u)\mathrm{d}u$$

where

- $x(u)$: Input signal as a function of time.
- $h(t-u)$: Impulse response of the system (a function that describes how the system responds to a very short input, like a "ping").

- The result of this integral is the output $y(t)$.

In simple terms:

- The system "convolves" the input $x(t)$ through its impulse response $h(t)$ to generate the output $y(t)$.

Working in the time domain is useful but can be very computationally intensive, especially for complex systems or signals. That's where the frequency domain comes in. Instead of looking at how signals behave over time, we analyse how they behave at different frequencies.

As explained before, Fourier transform can be used to convert signals from the time to the frequency domain. In the frequency domain, the convolution in the time domain becomes a simple multiplication:

$$Y(f) = X(f).H(f)$$

This equation says that at every frequency *f*, the system's output spectrum $Y(f)$ is the product of the input spectrum $X(f)$ and the system's frequency response $H(f)$. Therefore, instead of dealing with complex integrals in the time domain, the frequency domain simplifies things into straightforward multiplication.

Let's explore how the relationship between the input and output power spectra is derived. The power spectral density (PSD) describes how the power of a signal is distributed across different frequencies. If we want to find the power of a signal, we look at the magnitude squared of its spectrum. For example, the power at a frequency *f* in $Y(f)$ is proportional to $|Y(f)|^2$. Substituting this into the output power:

$$|Y(f)|^2 = |X(f).H(f)|^2$$

Using the properties of magnitudes, we can split this into:

$$|Y(f)|^2 = |X(f)|^2.|H(f)|^2$$

Power spectra represent the average power over time for random or continuous signals. Therefore:

- $G_{xx}(f) = |X(f)|^2$: Input power spectrum.
- $G_{yy}(f) = |Y(f)|^2$: Output power spectrum.

Substituting these into the equation, we get:

$$G_{yy}(f) = |H(f)|^2 .G_{xx}(f)$$

This formula explains how the power at each frequency in the output is related to the power in the input:

- $G_{xx}(f)$: Tells us how much power is present in the input signal at frequency *f*.
- $|H(f)|^2$: Describes how the system modifies that power at frequency *f*:
  - If $|H(f)|^2 > 1$, the system amplifies the input at that frequency.
  - If $|H(f)|^2 < 1$, the system attenuates the input at that frequency.
- $G_{yy}(f)$: Represents the resulting output power at frequency *f*, after the system has modified the input.

5.7.1 Estimation of H_1 for a Linear System

The H_1 estimator is a widely used method for estimating the FRF of a linear system. It is particularly effective in situations where noise predominantly affects the output signal, as shown in Figure 5.40. By leveraging mathematical tools such as spectral analysis and statistical expectations, the H_1 estimator isolates the contribution of the system dynamics from noise and input signals.

Consider a linear system with noise present only at the output:

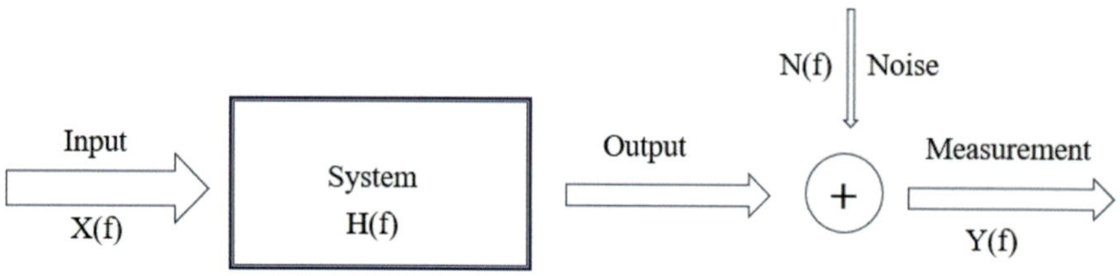

FIGURE 5.40 Linear system with noise in output signal.

$$Y(f) = X(f).H(f) + N(f)$$

in which $N(f)$ is the noise spectrum at the output, and the rest of the parameters are as defined before. The model assumes that the input $X(f)$ is deterministic or can be measured accurately, while $N(f)$ is stochastic and statistically uncorrelated with the input.

To derive the estimator, we multiply both sides of the system equation by the complex conjugate of the input signal, $X^*(f)$.

$$X^*(f).Y(f) = X^*(f).X(f).H(f) + X^*(f).N(f)$$

Taking the expected value of both sides gives:

$$E[X^*(f).Y(f)] = E[X^*(f).X(f).H(f) + X^*(f).N(f)]$$

where $E[.]$ represents the statistical expectation, which is commonly approximated using averages in practical applications. The expectations are then expressed in terms of spectral densities:

$$G_{yx}(f) = G_{xx}(f).H(f) + G_{nx}(f)$$

where:

- $G_{yx}(f)$ is the cross-spectral density between the output and input.
- $G_{xx}(f)$ the auto-spectral density of the input.
- $G_{nx}(f)$ the cross-spectral density between noise and input.

Spectral densities are estimated using techniques such as FFT.

As mentioned earlier, the H_1 estimator assumes that the input signal is uncorrelated with the output noise, leading to $G_{nx}(f) = 0$. This simplifies the equation to:

$$G_{yx}(f) = G_{xx}(f).H(f) \Rightarrow \hat{H}_1(f) = \frac{\hat{G}_{yx}(f)}{\hat{G}_{xx}(f)}$$

The H_1 estimator is robust against output noise but sensitive to input noise. If noise affects both the input and output, alternative estimators should be used. The H_1 estimator also assumes that the system behaves linearly. Nonlinear systems require alternative approaches.

5.7.2 Estimation of H_2 for a Linear System

The H_2 estimation method is commonly used for identifying the FRF of a linear system when noise is present only at the system's input. Consider a linear system with an input signal $X(f)$, a noise signal $M(f)$ present only at the input, and an output signal $Y(f)$. The system can be expressed in the frequency domain as:

$$Y(f)=\left[X(f)-M(f)\right].H(f)$$

To extract $H(f)$, multiply both sides by the complex conjugate of the output, $Y^*(f)$, and taking the average value, represented as the expected value operator, leads to:

$$G_{yy}(f)=[G_{xy}(f)-G_{my}(f)].H(f)$$

in which:

- $G_{yy}(f)=E\left[Y^*(f).Y(f)\right]$ is the auto-spectral density of the output,
- $G_{xy}(f)=E\left[X^*(f).Y(f)\right]$ is the cross-spectral density between the input and output,
- $G_{my}(f)=E\left[M^*(f).Y(f)\right]$ is the cross-spectral density between the noise and output.

Since the noise $M(f)$ is uncorrelated with the output $Y(f)$, $G_{my}(f)=0$. Substituting this condition and solving for $H(f)$ results in:

$$\hat{H}_2(f)=\frac{\hat{G}_{yy}(f)}{\hat{G}_{xy}(f)}$$

This equation forms the basis for H_2 estimation. Since $G_{yy}(f)$ is a real quantity, the phase angle of $\hat{H}_2(f)$ is derived from the cross-spectral density $G_{xy}(f)$. Furthermore, the relationship $G_{xy}(f)=G^*_{yx}(f)$ describes the reciprocity of the cross-spectral densities. Since $G_{yy}(f)$ is in the numerator of H_2, the H_2 estimate is greater than or equal to $H(f)$. However, since $G_{xx}(f)$ is in the denominator of H_1, the estimate is less than or equal to $H(f)$:

$$\left|\hat{H}_1(f)\right|\leq\left|H(f)\right|\leq\left|\hat{H}_2(f)\right|$$

5.7.3 Coherence Function

The coherence function provides an important measure for analysing the relationship between two signals in the frequency domain. The ratio of H_1 to H_2 is directly related to the coherence function, $\gamma^2(f)$, which is a measure of how well the input $x(t)$ and output $y(t)$ are correlated at a specific frequency. The coherence function is defined as:

$$\hat{\gamma}_{yx}{}^2(f) = \frac{\hat{H}_1(f)}{\hat{H}_2(f)} = \frac{\left|\hat{G}_{yx}(f)\right|^2}{\hat{G}_{xx}(f).\hat{G}_{yy}(f)}$$

If H_1 equals H_2, it implies a perfect correlation between the input and the output, with no noise affecting either signal. In such a scenario, the coherence function would equal 1 across all frequencies, indicating that the output is entirely generated by the input. Conversely, if H_1 and H_2 diverge, the coherence function will drop below 1, reflecting the presence of uncorrelated noise or nonlinear effects in the system.

For dynamic tests, the relationship between H_1, H_2, and the coherence function provides valuable insights into the quality of the measured FRF. Noise or inconsistencies in the input or output signals can reduce the coherence, affecting the accuracy of H_1 and H_2.

As an example, let's try to find H_1, H_2, and coherence function for Neka bridge. In this example, we will use Channel 5 (vertical accelerometer in southern truss, 6 m from the bearing) and Channel 12 (vertical accelerometer in southern truss, 18 m from the bearing). Channel 5 (CH05) is considered as the input signal (excitation) and channel 12 (CH12) is considered the output signal (response). Using MATLAB, the following procedure can be employed to compute the parameters of interest from the measured data. The results are presented in Figure 5.41.

```
a05 = resample(CH05, 1, 10); % Downsample input channel data
a12 = resample(CH12, 1, 10); % Downsample output channel data

% Cross-power spectral density (CPSD) and auto-power spec-
tral density (APSD)
[p12_5, fr] = cpsd(a12, a05, [], [], [], 200); % Cross-
spectrum between CH12 and CH05
[p5_5, fr] = cpsd(a05, a05, [], [], [], 200); % Auto-
spectrum of CH05
[p12_12, fr] = cpsd(a12, a12, [], [], [], 200); % Auto-
spectrum of CH12
[p5_12, fr] = cpsd(a05, a12, [], [], [], 200); % Cross-
spectrum between CH05 and CH12

% Estimation of H1 and H2
H1 = p12_5 ./ p5_5; % Frequency response using CH05 as input
H2 = p12_12 ./ p5_12; % Frequency response using CH12 as input
G = H1 ./ H2; % Coherence function estimation
```

```
% Plotting the frequency response functions
figure(1);
plot(fr(1:100), abs(H1(1:100)), 'r-', fr(1:100), abs(H2(1:100)), 'b-');
legend('H1', 'H2');
xlabel('Frequency (Hz)');
ylabel('Magnitude');
title('Frequency Response Functions (H1 and H2)');

% Plotting the coherence function
figure(2);
plot(fr(1:100), abs(G(1:100)));
xlabel('Frequency (Hz)');
ylabel('Coherence');
title('Coherence Function Estimation');
```

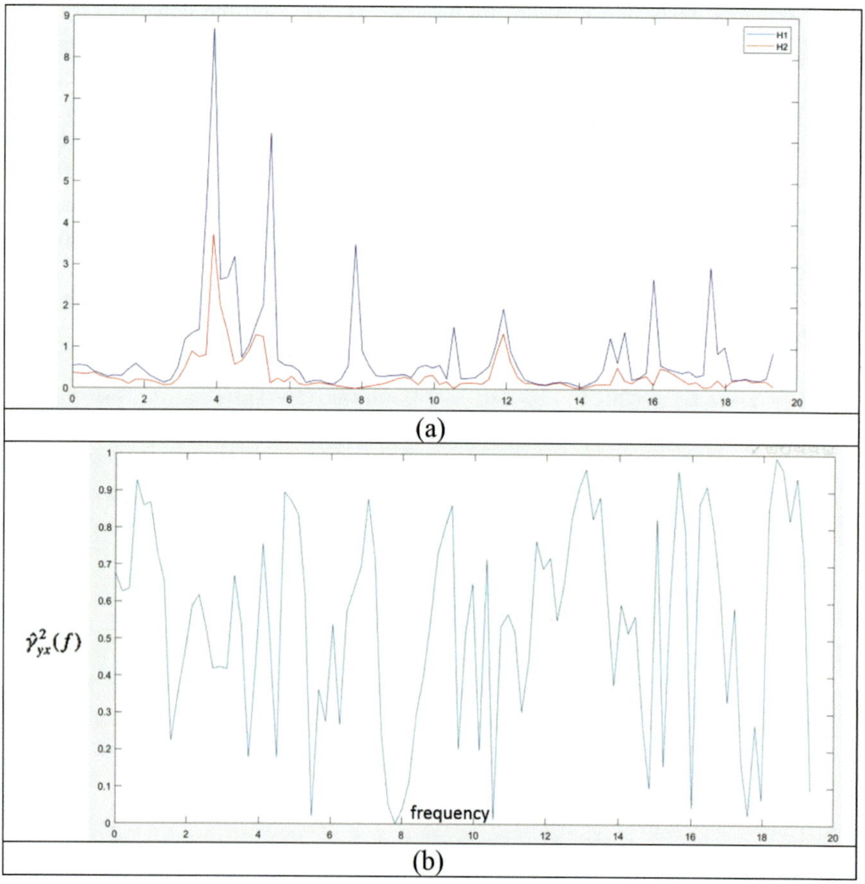

FIGURE 5.41 Estimation of (a) H_1 and H_2 and (b) coherence function using dynamic test results of Neka Bridge.

5.8 SUMMARY

In this chapter, the modal analysis process, including the determination of natural frequencies, damping, and mode shapes for a multi-degree-of-freedom structure, was first explained. A two-dimensional modal analysis of a truss bridge was presented as a case study.

Next, various structural vibration tests were described, such as impact excitation testing, shaker excitation testing, ambient vibration testing, and dynamic vehicle excitation testing. The experimental modal analysis process, which involves extracting the dynamic properties of a structure from measured data, was then discussed.

One of the simplest methods of experimental modal analysis is the representation of the frequency response using the free vibration portion of the measured acceleration data. In this context, the definition of the continuous Fourier transform and the procedure for obtaining the discrete Fourier transform were explained.

To reduce noise, the use of the power spectral density function for determining natural frequencies was presented. Challenges related to spectral leakage and the application of the Hanning window to address these issues were also discussed.

The experimental estimation of natural frequencies, mode shapes, and damping for the case study of the truss bridge was elaborated. Additionally, the issue of converting acceleration to displacement through double integration was introduced.

The frequency response function (FRF) of a system or structure was theoretically defined, and a frequency analysis of the truss bridge was presented as an example. Finally, the estimation of the frequency response function from measured data, considering noise in the input or output, was discussed.

REFERENCE

[1] A. Brandt, *Noise and Vibration Analysis*. Wiley, 2023.

6 Damage Diagnosis

6.1 INTRODUCTION

After measuring the data and preparing it through preprocessing and frequency analysis, structural health assessment or damage diagnosis becomes necessary. This process requires the definition, calculation, and classification of indices (into healthy, damaged, or various damage states) that are sensitive exclusively to damage.

In technical literature, numerous indices have been introduced as damage-sensitive metrics. These indices are analysed for their susceptibility to environmental conditions, such as temperature, humidity, and radiation, and normalization techniques are employed to eliminate the influence of factors unrelated to damages. Normalization ensures that the indices reflect only the structural damages and not external environmental effects.

If, due to factors such as insufficient data, it is not possible to classify damage-sensitive indices into normal and damaged states, alternative methods such as novelty detection or outlier analysis are typically employed. These methods identify anomalies or unusual data points that indicate potentially damaged conditions, serving as a precautionary measure when direct classification of damages cannot be achieved. These are summarized in Table 6.1.

6.2 FOUNDATIONAL PRINCIPLES OF DAMAGE DIAGNOSIS

Several principles guide the damage diagnosis process, ensuring accuracy and reliability in identifying structural issues, as shown in Figure 6.1 [1]:

1. *Inherent material defects*: Every material has micro-level flaws and cracks that can influence its performance. Recognizing these inherent defects is vital; damages must be quantitatively defined and measured based on the specific application. Understanding the typical defect patterns in a material type is crucial for establishing baseline status indicators.
2. *Comparison with baseline*: Damage assessment necessitates a comparative analysis between normal and damaged states. This principle underscores the importance of a reference sample or standard. For instance, in bridge monitoring, data collected from sensors on a newly constructed bridge can

DOI: 10.1201/9781003494652-6

TABLE 6.1
Diagnosis methods based on damage-sensitive indices

	Damage detection technique	
Damage-sensitive index	**Classification (Supervised Learning)**	**Outlier analysis (Unsupervised Learning)**
Scalar	Statistical methods	Statistical methods
Vector (1D)	Machine learning (MLP network)	Clustering
Image (2D)	Deep learning (convolutional network)	Encoder network

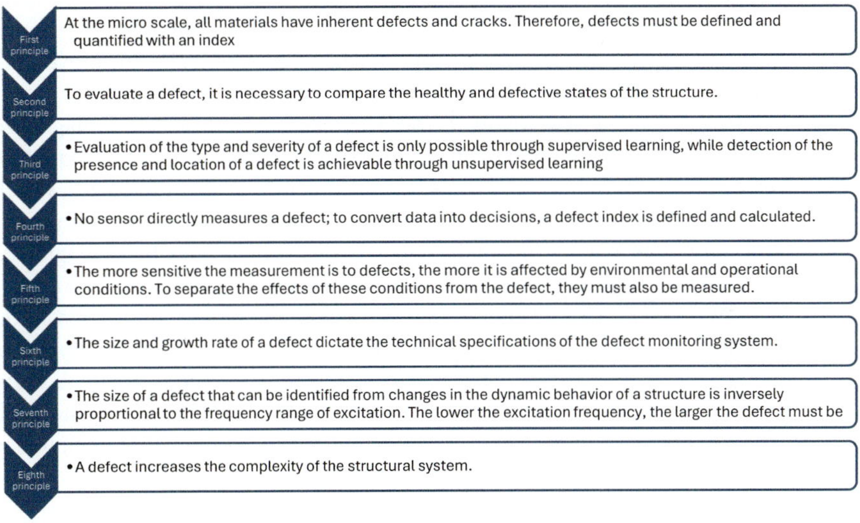

FIGURE 6.1 Foundational principles of damage diagnosis (adapted from [1]).

serve as a baseline for detecting anomalies in older bridges [2]. By identifying deviations from this baseline, engineers can pinpoint damages with greater accuracy.

3. *Damage diagnosis stages*: The stages of damage diagnosis encompass identification, location, type, and severity. Novelty detection techniques can ascertain the existence and location of damages, while supervised learning methods are essential for identifying damage types and severity. For instance, a case study involving the application of machine learning algorithms on highway bridges demonstrated how these methods could classify damage types based on historical data [3].
4. *Indirect measurement*: No sensor directly measures damages. Instead, indicators derived from sensor data are analysed to infer the presence of

damages. For example, strain gauges may not measure a crack directly but can indicate stress concentrations that suggest the underlying damages. This indirect measurement necessitates a thorough understanding of the relationship between measured indicators and potential damages.

5. *Sensitivity and environmental impact*: Damage indicators with high sensitivity are often more susceptible to environmental and operational conditions. Accurate diagnosis requires that both damage indicators and environmental factors be measured simultaneously. For example, studies on wind turbines have shown that environmental conditions can significantly affect vibration measurements, leading to false-positives in damage detection [4]. Thus, differentiating between environmental influences and actual damages is crucial.
6. *Damage growth rate and sensor specifications*: The rate of damage growth impacts the specifications of both sensors and data retrieval systems. Rapidly growing damages may require high-frequency sampling and more sensitive sensors to detect changes in real time. An example is the monitoring of fatigue cracks in steel structures, where timely detection is essential for preventing catastrophic failures [5].
7. *Dynamic behaviour and damage size*: The relationship between damage size and structural dynamic behaviour is inverse concerning excitation frequency. Lower excitation frequencies allow for the detection of larger damages, which is essential in applications such as earthquake engineering, where large structural shifts can occur [6]. Understanding this relationship helps engineers select appropriate diagnostic methods based on the expected damage sizes.
8. *Increased complexity*: Damages can introduce complexity into a structure, often transforming its behaviour from linear to nonlinear. This shift complicates damage diagnosis, as traditional linear models may no longer apply. A study on composite materials highlighted how the onset of delaminations alters the expected response under load, necessitating advanced modelling techniques for accurate diagnosis [7].

These principles form the foundation for effective damage diagnosis, ensuring that indicators used are reliable, accurate, and reflective of the actual condition of the structure or material. By adhering to these principles, engineers can implement robust diagnostic systems that enhance the safety and longevity of structures.

6.3 DEFINITION OF DAMAGE

Damage is defined as any alteration within a system that adversely affects its current or future performance. This high-level definition includes changes in material properties, structural geometry, boundary conditions, and system continuity. The implications of such damage can be profound, often leading to structural failures if not detected and addressed in a timely manner. Damage can manifest in two primary forms: suddenly (time-invariant) or through gradual, time-dependent processes.

Time-dependent damage accumulates incrementally over extended periods, commonly seen in phenomena such as fatigue or corrosion-related degradation.

For instance, fatigue damage in steel structures results from repeated loading and unloading cycles, which may lead to the formation of micro-cracks that gradually grow over time. A notable case is the collapse of the Silver Bridge in 1967, which was attributed to fatigue failure in a critical eye-bar that had undergone repeated stress cycles [8].

In contrast, sudden or time-invariant damage is often induced by discrete events such as earthquakes, severe weather conditions, or unexpected loads (e.g. heavy trucks). The sudden onset of damage can significantly compromise the integrity of a structure. For example, the 2010 Haiti earthquake caused extensive and immediate damage to buildings and infrastructure due to the rapid ground movement, illustrating how such events can induce structural damages [9].

Damage severity can be characterized geometrically (such as through crack geometry), by its effect on the load-bearing capacity of the structure (e.g., loss of stiffness or mass), or by alterations in the system's energy dissipation properties. For instance, a study on the seismic performance of reinforced concrete frames found that even small cracks can significantly reduce the energy dissipation capacity, which is crucial during seismic events [10]. Thus, damage can be defined as the partial or total degradation of the material structure, leading to a loss of strength in the affected structural components or the entire structure.

6.3.1 Common Causes of Damage

Damage to structures can arise from a variety of sources, primarily due to excessive loading from events such as earthquakes, accidents, storms, or heavy loads. Additionally, damage can stem from chemical deterioration (e.g., corrosion), mechanical degradation (e.g., cracking), physical degradation (e.g., freeze–thaw cycles or fire), or a combination of these factors.

Among these causes, corrosion in steel materials and cracking in steel or concrete materials are particularly prevalent. Corrosion can significantly weaken structural components, leading to costly repairs and potential safety hazards. Cracking, on the other hand, can occur due to various stressors, including thermal expansion and contraction, as well as loading conditions.

Understanding these common causes of damage is crucial for developing effective monitoring and maintenance strategies to prolong the lifespan of structures and ensure public safety.

6.4 DAMAGE-SENSITIVE INDICATORS

Damages induce detectable changes in a system, and the sensitivity of these changes to damages must be sufficient for detection amidst measurement errors and influencing parameters.

For instance, in SHM of bridges, changes in the natural frequency of a structure can indicate the presence of damage. A study by Zhang et al. [11] demonstrated that shifts in natural frequency correlated with increasing crack sizes, establishing a damage-sensitive indicator for assessing bridge integrity.

Another approach involves focusing on specific signal segments, like peak values. For example, in vibration monitoring of rotating machinery, peaks in vibration amplitude can indicate damage conditions, such as imbalance or misalignment. An example of this application is seen in the predictive maintenance strategies employed in wind turbine monitoring, where increased vibration levels can signal the need for maintenance before catastrophic failures occur [12].

Additionally, damage-sensitive indicators can be developed directly from signal analysis. Techniques such as wavelet transforms or machine learning algorithms can enhance the ability to detect subtle changes in signal characteristics that may indicate damages. These advanced methods have shown promising results in various applications, including the monitoring of civil infrastructure and mechanical systems.

6.4.1 Scalar Indicators

6.4.1.1 Statistical Characteristics of the Signal

Statistical indicators, such as the maximum, mean, root mean square (RMS), skewness, and kurtosis, serve as critical damage-sensitive indicators to differentiate normal states from various damaged states in structural monitoring. Each of these statistical measures provides unique insights into the behaviour of a system and can highlight changes that indicate potential damages.

- Maximum: The maximum value of a signal can identify extreme responses due to damage conditions. For example, in vibration monitoring of bridges, an increase in the maximum displacement response could suggest the onset of significant structural issues.
- Mean and RMS: The mean provides a central tendency of the signal, while the RMS reflects the average energy of the signal, which can help quantify the extent of deviations from normal operating conditions. A study by Zhang et al. (2016) demonstrated that RMS values increased significantly in a SHM system when cracks began to propagate in concrete beams, indicating a deterioration in structural integrity.
- Skewness and kurtosis: Skewness measures the asymmetry of the signal distribution, while kurtosis indicates the peakedness of the distribution. High kurtosis values may suggest that extreme events are occurring more frequently, which can be a sign of developing damages. Research has shown that changes in skewness and kurtosis can be effective indicators in detecting incipient damage in rotating machinery.

These statistical characteristics collectively enhance damage diagnosis by providing a robust framework for monitoring changes in structural behaviour over time, enabling proactive maintenance and intervention strategies.

6.4.1.2 Natural Frequency or Damping Indices

Natural frequency or damping, as estimated from field measurements, can be used as damage-sensitive indices. However, technical literature reports that the natural frequency of bridges is highly sensitive to environmental conditions, while damping is significantly influenced by the magnitude and method of excitation. These sensitivities make normalization—removing the effects unrelated to damages—a challenging task.

As mentioned, if it is possible to measure and estimate the natural frequency or damping values under both healthy and damaged conditions, or if theoretical models can predict their changes under damaged conditions, damage diagnosis can be conducted by classifying the indices into healthy and damaged states.

If measuring or estimating these indices under damaged conditions is not feasible, anomaly detection through outlier analysis can be employed. By identifying deviations or abnormal values in the data, potential damage conditions can be flagged for further investigation.

6.4.2 Vector Indicators

6.4.2.1 Signal Shape, Spectrum, or Correlation Comparison

The waveform shape of a signal is a high-dimensional feature that can serve as a damage-sensitive indicator, as changes in signal patterns can reveal underlying structural issues. By analysing the shape, spectrum, and correlation of signals, engineers can gain insights into the dynamic behaviour of structures under various loading conditions, such as traffic, wind, or seismic activity.

In structural health monitoring, signal shape analysis helps identify anomalies by comparing current signal patterns with reference or baseline signals. For instance, if a reference sample signal, as illustrated in Figure 6.2, differs significantly from the damaged sample signal, this discrepancy can indicate structural degradation. For real-world applications, researchers have analysed bridge structures where the shape of vibration signals effectively reflected localized damage in bridge girders [13]. Such studies confirm that shape changes in vibration signals are highly indicative of structural issues, making this approach beneficial for proactive maintenance.

The spectrum of a signal, represented in Figure 6.3, provides additional damage-sensitive information. Spectrum analysis, which involves transforming the signal into the frequency domain, enables the identification of frequency components associated with specific types of structural anomalies. For example, when cracks form in steel beams, they often alter the resonant frequencies. By analysing spectral changes over time, engineers can identify these damage patterns and forecast potential failure points [14]. Spectral changes are particularly useful in monitoring large structures like skyscrapers and bridges, where environmental factors constantly impact structural integrity.

Correlation analysis can further validate signal anomalies by assessing the similarity between signals collected at different times. Comparing autocorrelation patterns of reference and damaged signals, as shown in Figure 6.4, can reveal gradual or sudden changes that may indicate damage progression. In the field, correlation

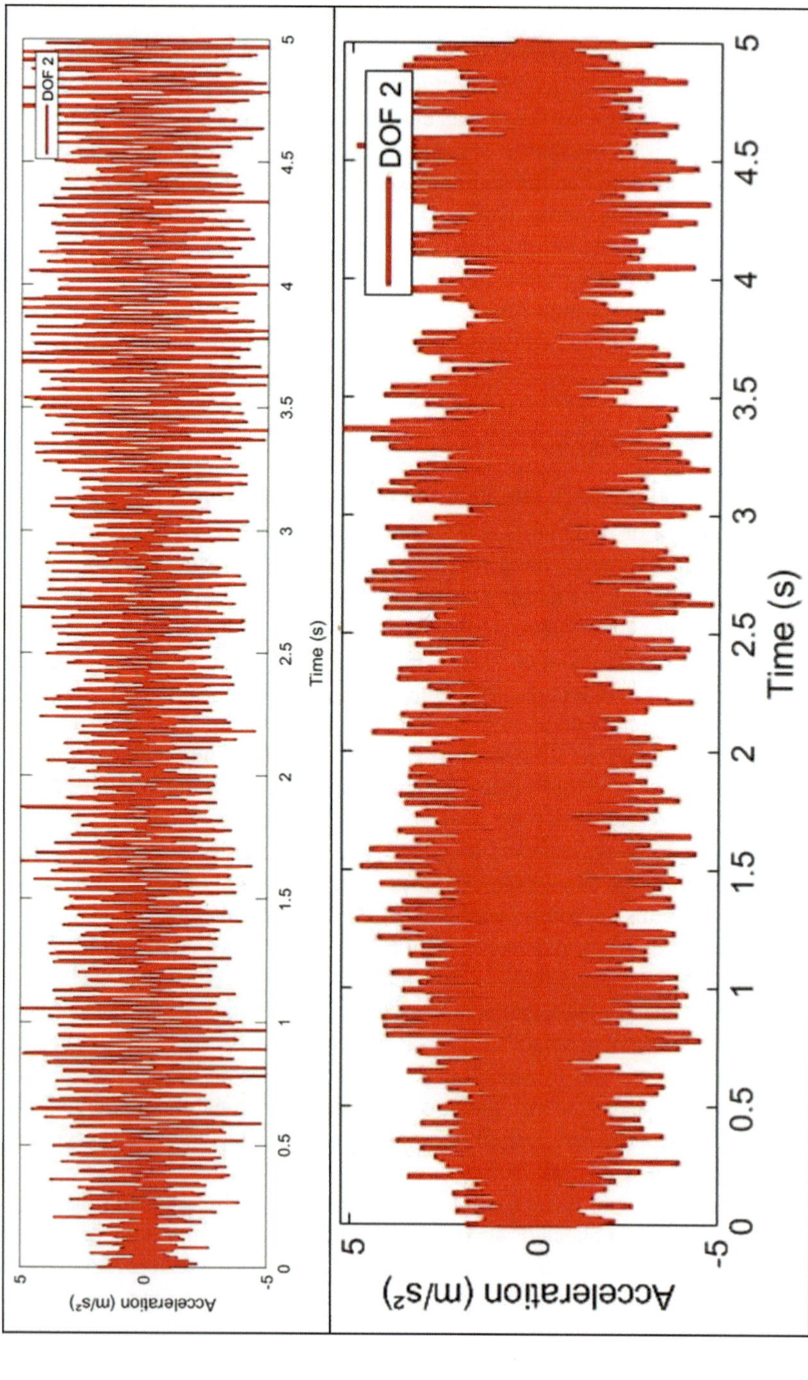

FIGURE 6.2 Comparison between reference sample signal and damaged sample signal of 3DOF model (derived from 30% reduction at element 26 of 50 DOF model of Neka bridge).

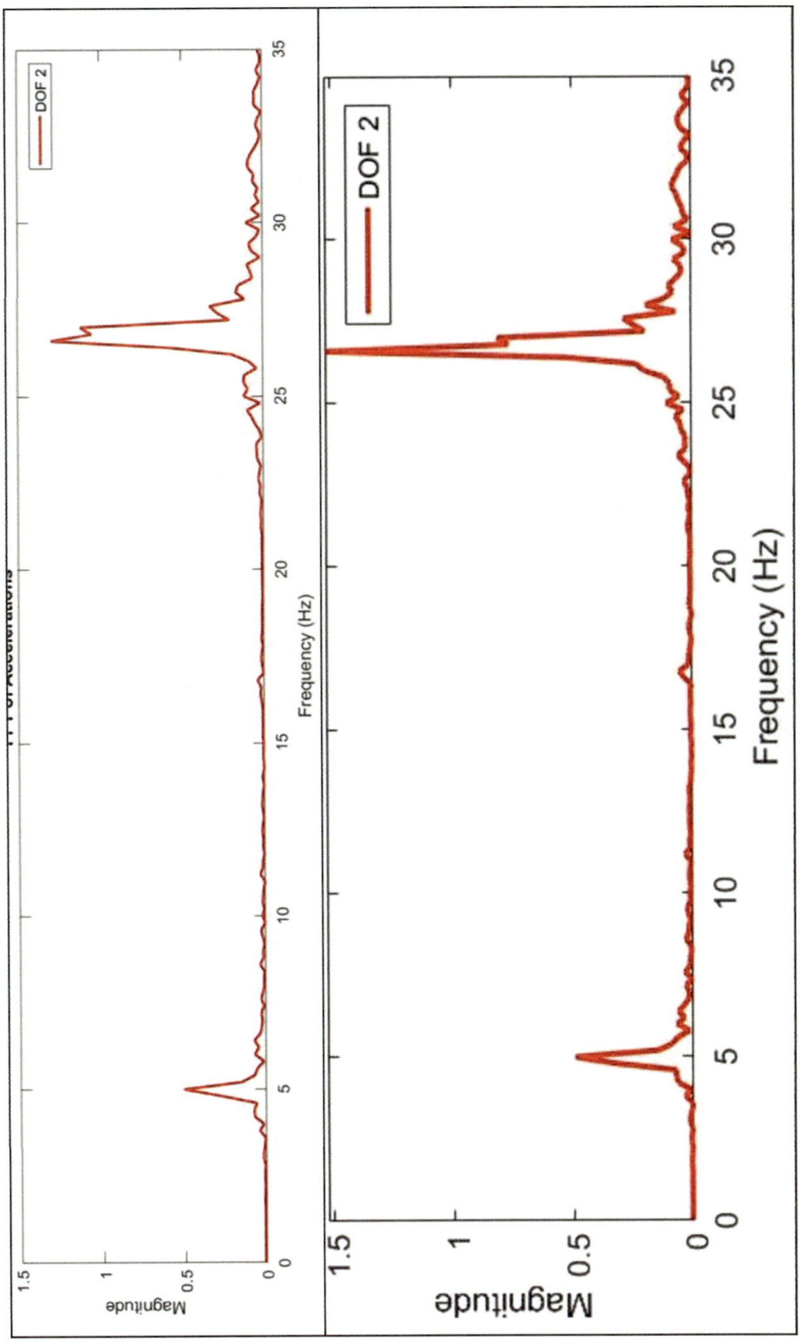

FIGURE 6.3 Comparison of the spectrum of the reference sample signal with that of the damaged sample signal.

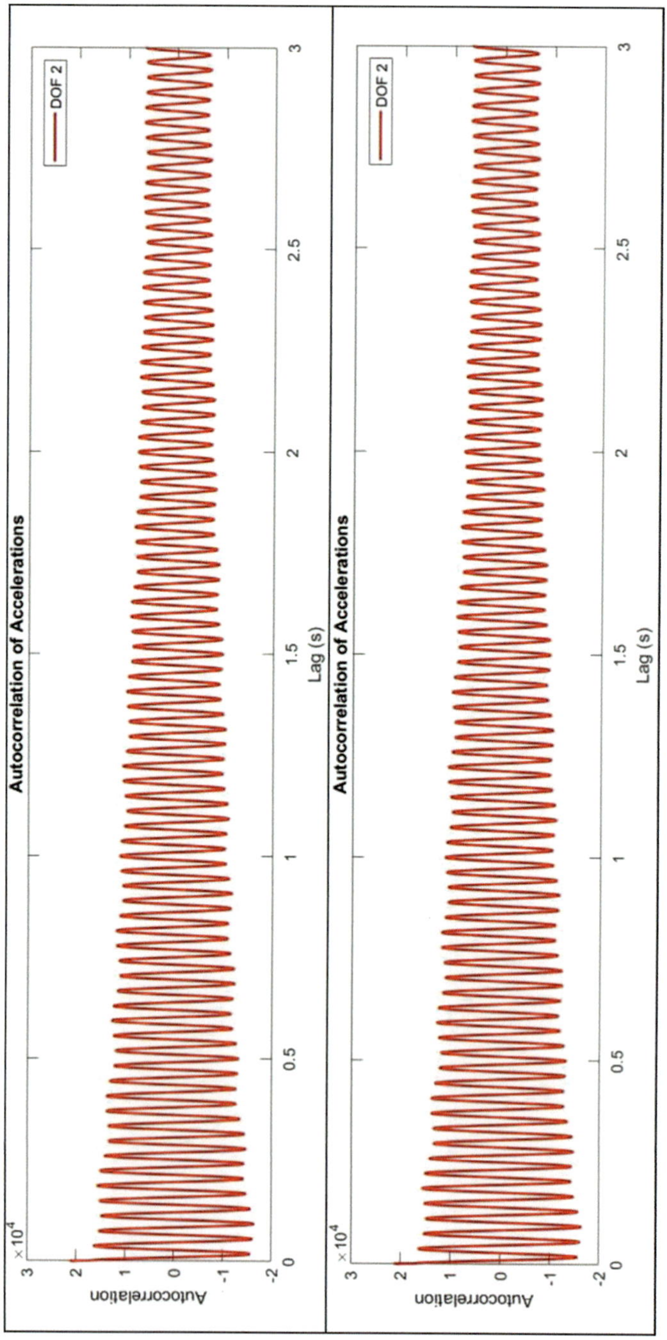

FIGURE 6.4 Comparison of autocorrelation for the reference sample signal with that of the damaged sample signal.

techniques have been successfully applied to monitor railway tracks and detect minor track misalignments that, if left unchecked, could evolve into major structural problems [15].

To enhance signal analysis, damage-sensitive indicators are often derived through advanced computational techniques, including wavelet analysis and machine learning algorithms. Wavelet analysis, for example, decomposes signals into multiple frequency bands, making it possible to detect transient events or localized damages that traditional Fourier analysis might overlook. A study by Chakraborty et al. [16] demonstrated that wavelet-based feature extraction effectively highlighted damage in steel-reinforced concrete structures, significantly improving damage detection rates.

6.4.2.2 Mode Shapes

Comparing the mode shapes of a structure in its normal (undamaged) and damaged states can reveal significant insights into the presence and location of structural damage. Mode shapes are patterns of free vibration that a structure exhibits at specific frequencies, and deviations in these patterns can indicate damage. Several metrics are used to compare these mode shapes, including the modal assurance criterion (MAC), the coordinate modal assurance criterion (COMAC), and the Yuan criterion. Each of these metrics focuses on different aspects of mode shape analysis to detect and locate structural anomalies.

6.4.2.2.1 Modal Assurance Criterion

The modal assurance criterion (MAC) is a widely used metric to quantify the similarity between mode shapes in two different states of a structure—typically the undamaged (reference) and damaged (current) states. Mathematically, MAC is defined as:

$$\mathrm{MAC}\left(\{\psi\}_u,\{\psi\}_d\right)=\frac{\left|\{\psi\}_u^T\cdot\{\psi^*\}_d\right|^2}{\left(\{\psi\}_u^T\cdot\{\psi^*\}_u\right)\left(\{\psi\}_d^T\cdot\{\psi^*\}_d\right)}$$

where

- ψ_u is the mode shape vector in the undamaged state,
- ψ_d the mode shape vector in the damaged state,
- T denotes the transpose operation,
- * denotes the complex conjugate (used in cases of complex mode shapes).

The MAC value ranges between 0 and 1, where a value close to 1 indicates high similarity between the mode shapes, while a value close to 0 suggests a lack of correlation, which could be due to structural changes or damage. By calculating the MAC values between mode shapes in different states, engineers can identify changes in vibration characteristics, which may indicate damage locations.

Considering the 50 DOF model of Neka bridge, modulus of elasticity of element 26 is reduced by 30% to simulate a damage in this member. Figure 6.5 shows MAC

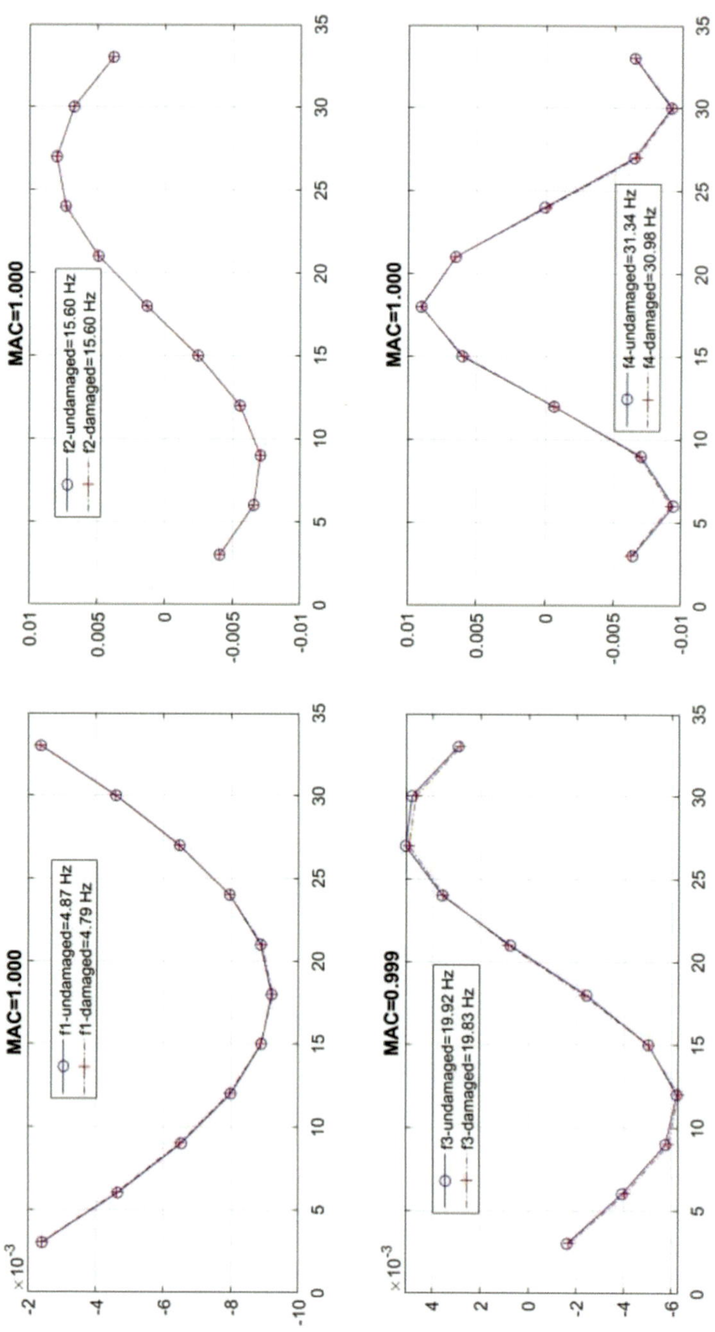

FIGURE 6.5 Change in mode shape due to a simulated damage in element 26 of 50 DOF model sof Neka Bridge.

for first four mode shapes of Neka bridge with and without the simulated damage. It is evident that the impact of damage was minimal on the structure.

6.4.2.2.2 Coordinate Modal Assurance Criterion

While MAC compares overall mode shapes, the coordinate modal assurance criterion (COMAC) offers a more localized approach by evaluating mode shape consistency at individual points, such as specific nodes in a finite element model. COMAC is particularly useful for identifying localized damages, as it highlights discrepancies in mode shapes at each coordinate.

For a given node i, and a set of N_m mode shapes, COMAC is defined as:

$$\mathrm{COMAC}(i) = \frac{\left(\sum_{j=1}^{N_m} \left|\psi_{\mathrm{u}_{i,j}} \cdot \psi_{d_{i,j}}\right|\right)^2}{\left(\sum_{j=1}^{N_m} \left|\psi_{\mathrm{u}_{i,j}}\right|^2\right)\left(\sum_{j=1}^{N_m} \left|\psi_{d_{i,j}}\right|^2\right)}$$

where

- $\psi_{u_{(i,j)}}$ is the mode shape amplitude for mode j at node i in the undamaged state,
- $\psi_{d_{(i,j)}}$ is the mode shape amplitude for mode j at node i in the damaged state.

COMAC values range from 0 to 1, similar to MAC, with values near 1 indicating consistent mode shape behaviour at a given node across the compared states. Nodes with lower COMAC values may be indicative of localized structural damage.

6.4.2.2.3 Mode Shape Curvature as a Damage-Sensitive Index

The curvature of a mode shape can also act as a sensitive indicator of structural damage, particularly in cases of local damages such as cracks. Mode shape curvature reflects the deformation pattern of a structure and tends to be highly sensitive to stiffness changes. In damaged regions, curvature variations are more pronounced, as structural damage often leads to localized stiffness reductions, which in turn affect the mode shape curvatures. Engineers can compute these curvatures and monitor them over time to detect early signs of structural distress.

To do so, it is needed to calculate the second derivative of modal shapes, as shown below:

$$\ddot{V}(\psi_i) = \frac{\psi_{i-1} - 2\psi_i + \psi_{i+1}}{h^2}$$

in which ψ_i is the mode shape value at a specific mode i, and h is the distance between locations whose mode shape has been calculated in it. For Neka bridge, results are presented in Figure 6.6.

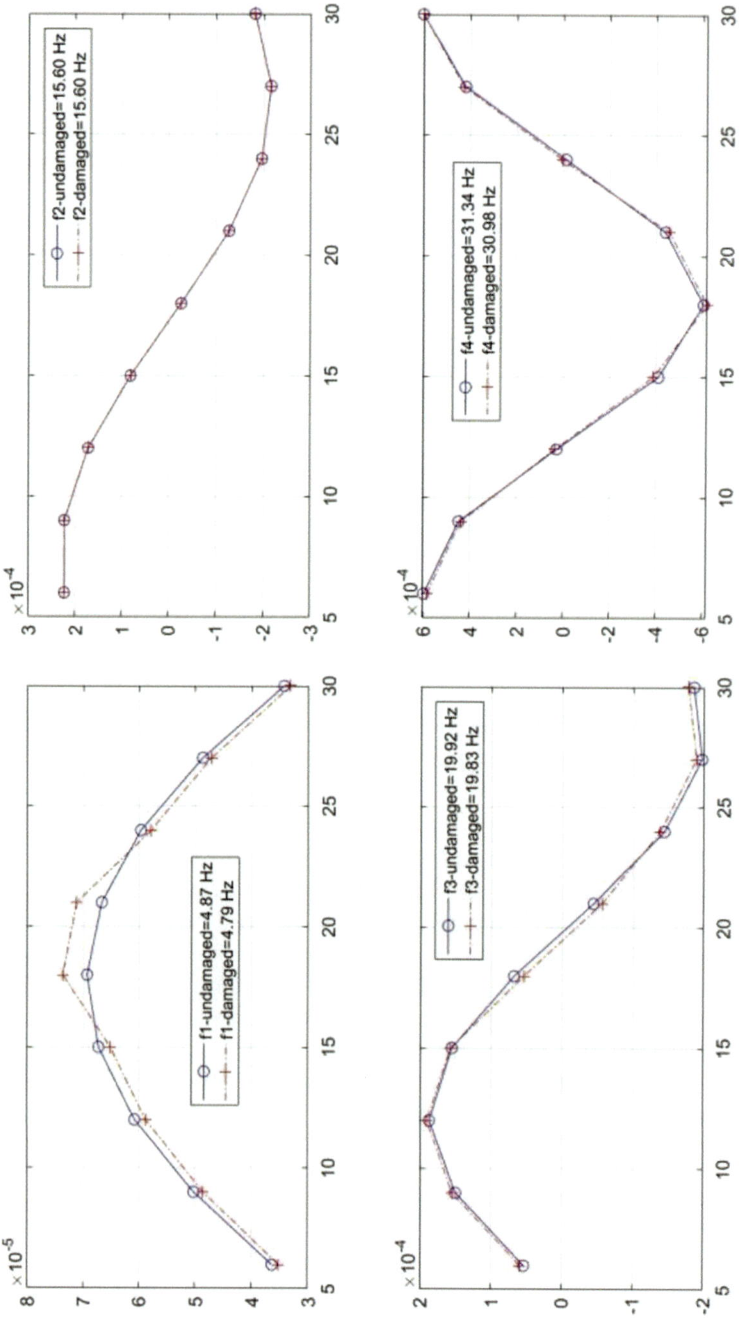

FIGURE 6.6 Mode shape curvature for Neka bridge.

6.4.2.2.4 Yuan Criterion

The Yuan criterion allows for the comparison of both the mode shape and the natural frequency of a structure in its normal and damaged states, providing an integrated view of structural changes due to damage. The criterion is formulated as:

$$\{Y\}_i = \frac{\{\psi\}_{i_d}}{\omega_{i_d}} - \frac{\{\psi\}_{i_u}}{\omega_{i_u}}$$

where

- $\{\psi\}_{i_d}$ and $\{\psi\}_{i_u}$ are the mode shape values at a specific mode i in the damaged and undamaged states, respectively,
- ω_{i_d} and ω_{i_u} are the natural frequencies for mode i in the damaged and undamaged states, respectively.

The Yuan criterion provides a ratio of mode shape changes relative to their respective frequencies, which can help in distinguishing between frequency shifts caused by general structural alterations versus those due to localized damage. If significant discrepancies are observed in Y_i across different modes, this could indicate areas where structural integrity has been compromised.

By combining these techniques—MAC for overall shape comparison, COMAC for localized assessment at nodes, mode shape curvature for sensitive damage detection, and the Yuan criterion for frequency-informed shape changes—engineers can comprehensively assess structural health. These methods support the early identification of damage, allowing timely interventions to maintain structural integrity and safety.

Using the simulated damage in element 26 from the previous section, Yuan criterion is calculated and is maximum for this element as shown in Figure 6.7.

In a system with more than one degree of freedom, there are multiple natural frequencies, damping ratio, and modal shapes, where all of them can be arranged into a vector. This approach allows the analysis of variations in the vector along the length of the bridge, providing a basis for damage diagnosis.

Furthermore, monitoring frequency behaviour patterns can help distinguish between normal and damaged states.

6.4.2.3 Principal Component Analysis

Principal component analysis (PCA) is a linear dimensionality reduction technique that transforms high-dimensional data into a set of uncorrelated variables called principal components, ordered by their ability to explain variance in the original dataset. By distilling complex measurements into a few summary indices, PCA serves as a powerful tool for identifying structural changes, making it particularly valuable in damage detection applications.

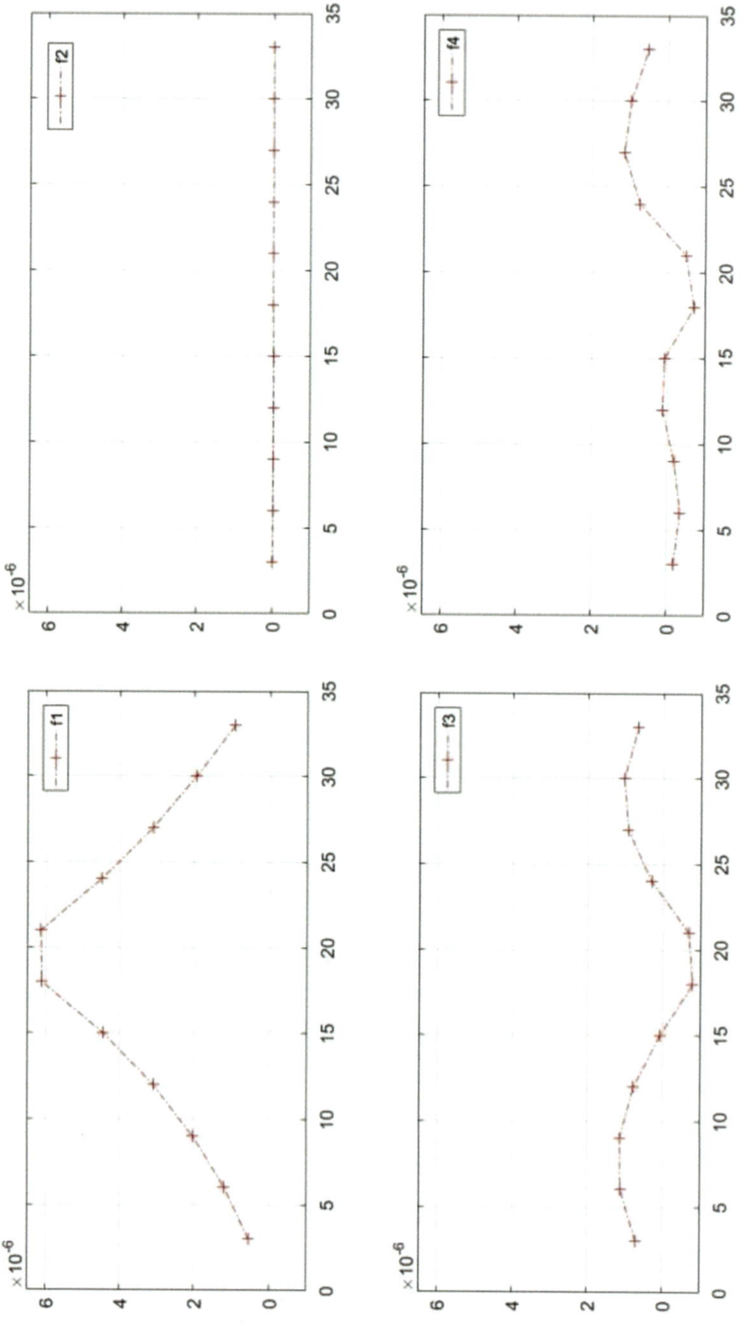

FIGURE 6.7 Yuan criterion for simulated damage in 50 DOF Model of Neka Bridge.

The transformation matrix in PCA is derived by solving:

$$C = X^{\mathrm{T}} X$$

where X is the matrix of data, and the eigenvectors of C represent the principal components.

X: A matrix, often representing a dataset where rows are observations and columns are variables.
X^{T}: The transpose of X—rows become columns and columns become rows.
$X^{\mathrm{T}} X$: The result of multiplying the transpose of X by X.

If X is centred (mean subtracted), then $X^{\mathrm{T}} X$ is proportional to the covariance matrix, which tells you how variables relate to each other.

Also, we can use Hankel matrix of data instead of $X^{\mathrm{T}} X$ and applying PCA on the Hankel matrix. A Hankel matrix is a special kind of structured matrix where each ascending skew-diagonal from left to right is constant. That is, for a matrix H, the element $H_{i,j}$ depends only on $i + j$. Formally:

$$H = [x_1 \; x_2 \; x_3 \; \cdots \; x_n$$

$$x_2 \; x_3 \; x_4 \; \cdots \; x_{n+1}$$

$$x_3 \; x_4 \; x_5 \; \cdots \; x_{n+2}$$

$$\vdots \; \vdots \; \vdots \; \ddots \; \vdots$$

$$x_m \; x_{m+1} \; x_{m+2} \; \cdots \; x_{m+n-1}$$

This structure is particularly useful when analysing time-series data, where each row of the matrix represents a lagged version of the original signal. By embedding the signal into a Hankel matrix, we can capture temporal correlations and apply PCA to extract dominant patterns or trends.

6.4.2.4 t-Distributed Stochastic Neighbour Embedding (t-SNE)

T-SNE is a nonlinear dimensionality reduction technique that specializes in visualizing complex, high-dimensional data by preserving local relationships between points. Unlike linear methods such as PCA, which prioritize global variance, T-SNE operates through a probabilistic framework that mimics how data points "choose neighbours" in both the original high-dimensional space and the reduced low-dimensional representation. This approach makes it particularly effective for uncovering clusters, manifolds, and subtle patterns that linear techniques might overlook.

t-SNE operates in two phases:

1. High-dimensional similarity modelling:
   - Computes conditional probabilities $P_{j|i}$ between data points using a Gaussian kernel, where closer points receive higher probabilities.
   - Symmetrizes probabilities to $p_{ij} = \frac{P_{i|j} + P_{j|i}}{2n}$, ensuring similar points attract each other.
2. Low-dimensional embedding:
   - Initializes random 2D/3D points and calculates similarities q_{ij} using a heavy-tailed Student's t-distribution to prevent overcrowding.
   - Minimizes Kullback–Leibler divergence $KL(P \parallel Q) = \sum_{i \neq j} p_{ij} \log \frac{p_{ij}}{q_{ij}}$ via gradient descent, aligning low-dimensional distances with original similarities.

One of T-SNE's defining characteristics is its emphasis on local structure preservation. While PCA might stretch or compress global patterns to maximize variance, T-SNE prioritizes maintaining the integrity of small-scale groupings, making it ideal for tasks like comparing signal shapes or analysing mode vectors in vibration studies. For instance, when visualizing spectral correlations or time-frequency features, T-SNE can reveal hidden clusters corresponding to distinct operational states or fault conditions that linear projections might conflate. However, this local focus comes with trade-offs: the axes in T-SNE plots lack intrinsic meaning, inter-cluster distances shouldn't be overinterpreted, and results vary slightly between runs due to the stochastic initialization process.

In practice, T-SNE requires careful parameter tuning. The perplexity parameter acts as a balancing knob—low values force tight, hyperlocal groupings, while higher values encourage broader pattern recognition. Many workflows first apply PCA to reduce dimensions to a manageable number (e.g., 50 components) before feeding the data into T-SNE, as this preprocessing step often improves both computational efficiency and result stability. Despite its limitations in interpretability and reproducibility, T-SNE remains indispensable for exploratory analysis of vector indices derived from signal spectra, correlation matrices, or mode shape decompositions, serving as a bridge between raw high-dimensional data and human intuition. When combined with PCA in a hybrid approach, it offers complementary insights—PCA for global feature extraction and T-SNE for detailed cluster topography—forming a powerful toolkit for modern signal processing challenges.

6.4.3 Image-Based Indicators

An image is a two-dimensional feature that, with advancements in artificial intelligence and deep learning methods, can be employed as a two-dimensional damage-sensitive index. These images may include time-frequency diagrams of vibration signals, actual photographs of structural components, or any two-dimensional images obtained through methods such as thermography.

Damage diagnosis can be performed by comparing the behaviour of similar components, extracting the healthy pattern, and identifying deviations from this healthy pattern. Such image-based indices are particularly powerful due to their ability to capture complex, spatial, and temporal relationships, enabling a more detailed analysis of structural conditions.

6.4.4 Numerical Model Updating

To identify the existence, location, and severity of damage, numerical model updating is performed by minimizing the differences between the numerical model predictions and actual field measurements. This process involves adjusting structural parameters such as stiffness, mass, internal forces, and boundary conditions (e.g., translational and torsional spring stiffness) to create a validated model that aligns closely with the field data.

For instance, Setsobhonkul et al. [17] conducted an investigation on a railway bridge, where they found significant discrepancies between the measured and predicted stiffness values. These discrepancies pointed to an incomplete or inaccurate representation of the bridge's true structural behaviour in the initial model. The differences between the predicted and observed data suggested that factors like ageing, wear, or potential damage had not been properly captured by the original model.

To address these mismatches, the researchers undertook a process of model updating, adjusting key structural parameters such as the modulus of elasticity, stiffness, and boundary conditions. They employed an iterative approach, using field data from sensors placed on the bridge to refine their predictions. As part of this process, they adjusted the bridge's translational and torsional spring stiffness to better align with the measured data. This series of adjustments led to a more accurate representation of the bridge's actual condition, ultimately improving the model's reliability.

The updated model, with its refined parameters, proved invaluable for diagnosing potential damages within the structure. For example, by pinpointing areas where the stiffness was compromised, the model helped engineers identify critical zones that were vulnerable to damage. The study highlighted the importance of numerical models in damage diagnosis, demonstrating that even small discrepancies between predicted and field-measured data can reveal underlying structural issues. These findings underscore the value of numerical model validation as a tool for proactive maintenance and accurate damage detection, ensuring the long-term safety and performance of infrastructure.

6.4.4.1 Numerical Model Validation Algorithm

Updating a numerical model involves adjusting model parameters within an appropriate range to better match observed data, which can be framed as an optimization problem. This process often involves defining an objective function, which quantifies the difference between simulated model results (such as natural frequencies or mode shapes) and actual measured data. The goal is to minimize this difference, thereby aligning the model more closely with the real structure's behaviour.

Optimization Problem Framing and Objective Function

Let's denote the parameters of the model as a vector *x*. These parameters could include physical properties such as material stiffness, damping coefficients, or boundary conditions. The optimization problem can be expressed as:

$$\min f(x) = \|\text{Model}(x) - \text{Measured Data}\|^2$$

where

- *f*(*x*) is the objective function, representing the difference between the model's predictions and the measured data,
- Model(*x*) represents the response of the numerical model given parameters *x*,
- Measured data is the set of actual data collected from the physical structure.

This optimization process seeks a solution x^* that minimizes *f*(*x*), ensuring that the model's output aligns as closely as possible with the measured data.

Sensitivity-Based Iterative Optimization Approach

One common way to approach this optimization problem is through the sensitivity-based iterative approach, a gradient-driven method that requires calculating the gradient of the objective function with respect to the optimization variables. This process is computationally intensive because each variable's impact must be evaluated individually, which involves recalculating the model response for each modified parameter.

The sensitivity matrix *S* represents how changes in each parameter affect the model's outputs, and it is defined as:

$$S = \frac{\partial \text{Model}(x)}{\partial x}$$

If there are *n* parameters to optimize, this matrix requires *n* model analyses, each with a different parameter perturbation. The sensitivity matrix is then used to estimate how adjusting each parameter will impact the objective function, guiding the iterative updates to *x*.

Iterative Optimization Process

1. *Initialization*: Start with an initial guess x_0 for the parameters.
2. *Objective Function Evaluation*: Calculate $f(x_0)$ to assess the initial discrepancy between model predictions and measured data.
3. *Sensitivity analysis*: Compute the sensitivity matrix S by conducting a series of model runs, each slightly adjusting one of the parameters.
4. *Parameter update*: Adjust parameters based on the sensitivity matrix and the gradient of the objective function. A common update rule is:

$$x_{k+1} = x_k - \alpha S^{T} \nabla f(x_k)$$

where

- α is a step size (learning rate) that controls the update magnitude,
- S^T is the transpose of the sensitivity matrix,
- $\nabla f(x_k)$ is the gradient of the objective function at the current parameter values.

5. Convergence check: If $f(x_{k+1})$ is within an acceptable range of tolerance, the optimization stops; otherwise, repeat steps 2–4.

Practical Implementation
This optimization process is commonly implemented by linking numerical analysis software (e.g., finite element analysis software for structural simulations) with optimization software that handles the iterative updates based on the sensitivity analysis. Each iteration involves recalculating the sensitivity matrix and updating the parameters, which can be computationally demanding. Advanced implementations might use parallel processing to reduce computation time, especially when the model requires extensive simulations.

Example Objective Functions
In structural health monitoring, common objective functions include:

1. Natural frequency matching:

$$f(x) = \sum_{i=1}^{N} \left(\frac{\omega_{i,\text{ measured}} - \omega_{i,\text{model}}(x)}{\omega_{i,\text{ measured}}} \right)^2$$

where $\omega_{i,\text{measured}}$ and $\omega_{i,\text{ model}}(x)$ are the i-th natural frequencies of the measured and modelled structures, respectively.

2. *Mode Shape Correlation*: If mode shapes are normalized, the MAC can be used within the objective function to minimize differences in shape between measured and modelled modes.

By refining the parameters iteratively through this sensitivity-based optimization, the model validation process achieves an optimal match with the actual structure's behaviour, enhancing the model's accuracy and reliability in predicting real-world structural responses.

Example: Consider a truss with two members subjected to a force of -2 kips applied at the second degree of freedom [18]. The initial cross-sectional area of the members was 1 in², but due to corrosion, this has changed. If the measured strain in members T1 and T2 are:

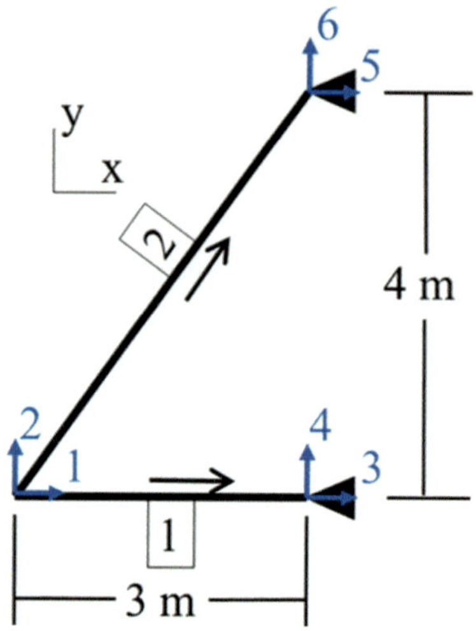

FIGURE 6.8 Two-member truss.

$$T1 = -625/(E \times 333)$$
$$T2 = 25/(E \times 25)$$

estimate the cross-sectional areas of the members based on the experimental data.

For this truss example with two members, the problem involves calculating the cross-sectional areas A_1 and A_2 using strain measurements and the optimization approach to fit the experimental data. Figure 6.8 shows a representation of the problem. Here's a breakdown of the optimization setup and results:

1. *Assumptions*:
   - q1 and q2 are the axial forces in the two truss members.
   - The analytical strains in each member are:
     - s1=q1/(E.A1)
     - s2=q2/(E.A2)
     - where E is the elastic modulus.
2. *Objective function*:
   - The objective function, *f*, is formulated to minimize the difference between measured strains T1 and T2 and analytical strains s1 and s2:

$$f=(T1-s1)^2+(T2-s2)^2$$

3. *Solution method*:

   The optimization problem is solved using Newton's method to iteratively adjust A1 and A2 for convergence.

   After several iterations, the method converges to: A1=0.8 in^2 and A2=0.7 in^2. values in each step are shown in Table 6.2

Force vector, stiffness matrix, and displacement vector of the two-member truss in the global coordinate system are as follows ({Q}=[K]×{D}):

$$\begin{bmatrix} 0 \\ -2 \\ Q_3 \\ Q_4 \\ Q_5 \\ Q_6 \end{bmatrix} =$$

$$E \times \begin{bmatrix} 0.333\times A_1 + 0.072\times A_2 & 0.096\times A_2 & -0.333\times A_1 & 0 & -0.072\times A_2 & -0.096\times A_2 \\ 0.096\times A_2 & 0.128\times A_2 & 0 & 0 & -0.096\times A_2 & -0.128\times A_2 \\ -0.333\times A_1 & 0 & 0.333\times A_1 & 0 & 0 & 0 \\ 0 & 0 & 0 & 0 & 0 & 0 \\ -0.072\times A_2 & -0.096\times A_2 & 0 & 0 & 0.072\times A_2 & 0.096\times A_2 \\ -0.096\times A_2 & -0.128\times A_2 & 0 & 0 & 0.096\times A_2 & 0.128\times A_2 \end{bmatrix}$$

$$\times \begin{bmatrix} D_1 \\ D_2 \\ 0 \\ 0 \\ 0 \\ 0 \end{bmatrix}$$

Force vector, stiffness matrix, and displacement vector of a truss member in the local coordinate system can be determined based on the following:

$$q = \frac{AE}{L}\begin{bmatrix} -\lambda_x & -\lambda_y & \lambda_x & \lambda_y \end{bmatrix} \begin{bmatrix} D_{Nx} \\ D_{Ny} \\ D_{Fx} \\ D_{Fy} \end{bmatrix}$$

D's are global joint displacements and the subscripts N and F stand for the near and far ends of the member. **λ's** are direction cosine of member. The above equations are then solved using the Newton method, and results for each step are presented in Table 6.2.

TABLE 6.2
Steps of Newton's method

Step	A1	A2	$F(A1, A2) \times 10^{-15}$
0	1.000	1.000	1,532,540
1	0.938	0.625	308,502
2	0.692	0.680	115,026
3	0.766	0.698	8567
4	0.796	0.700	111
5	0.800	0.700	0
6	0.800	0.700	0

6.4.4.2 Case Study of Numerical Model Updating for Neka Bridge

The location of installed strain gauges on the Neka truss bridge are shown in Figure 6.9. Also, Figure 6.10 shows the location of axles of a single locomotive on the span during the field measurements and the applied loads on truss nodes. Table 6.3 compares the measured versus numerical results at locations where strain was measured during the field testing.

It can be seen that the measured forces differ slightly from those estimated through numerical model. The following function is defined for updating the numerical model:

```
function [f] = TargetFunction2 (A)
force = [1; 26.5; 24.5; 3.5; 22.6; 27.5; 5.4];
Sforce = force .* [A(21:27)]‘; % South truss force
Nforce = force .* [1 - A(21:27)]‘; % North truss force
k = [
    -0.09, -0.19, -0.28, -0.37, -0.45, -0.65, -0.75;
     0.33, 0.56, 0.89, 1.18, 1.41, 2.08, 2.39;
    -0.28, -0.67, -1.02, -1.39, -1.62, -2.38, -2.73;
     0.49, 0.91, 1.4, 1.85, 2.22, 2.78, 2.18;
     0.09, 0.19, 0.28, 0.37, 0.45, -0.47, -0.37;
    -0.09, -0.19, -0.28, -0.37, 0.45, 0.47, 0.37;
    -0.44, -1.01, -1.53, -2.06, -2.43, -2.57, -2.06;
     0.65, 1.25, 1.91, 2.52, 1.82, 1.93, 1.51;
    -0.6, -1.35, -2.04, -2.73, -1.62, -1.71, -1.39;
     0.81, 1.6, 1.91, 1.71, 1.01, 1.07, 0.84
]; % stiffness matrix

test=[-184,-169;215,184;-286,-263;299,262;17,22;51,54;-362,-359;292,238;
-364,-304;209,197];%measured strains
Skf=k*Sforce; % south truss
Nkf=k*Nforce; %north truss
f=sum(([A(1:10)]‘.*test(:,1)*0.002-Skf).^2+([A(11:20)]’.*test(:,2)*0.002-
Nkf).^2);%target function
```

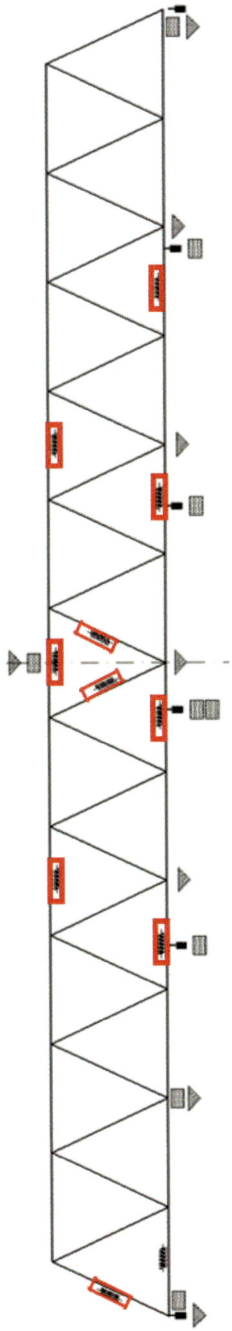

FIGURE 6.9 Strain measurement locations in the northern and southern trusses of the Neka bridge.

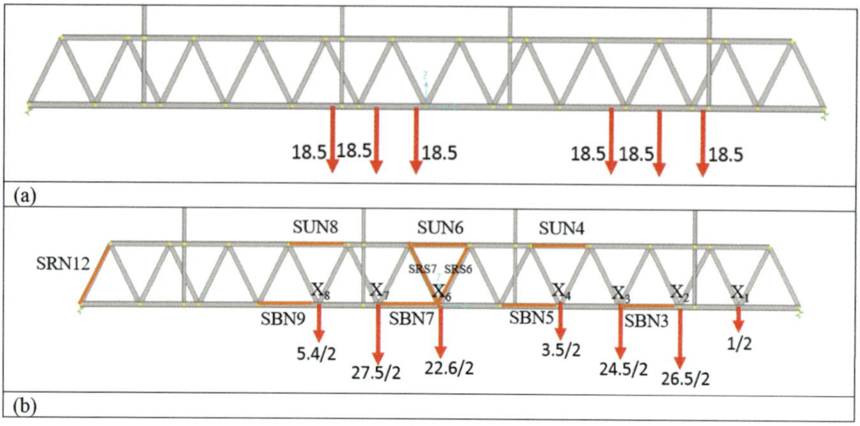

FIGURE 6.10 (a) Location of locomotive axles during the field testing and (b) applied load on the truss nodes under single locomotive loading during the field testing.

TABLE 6.3
Comparison of forces from field measurement versus numerical model

Southern truss	Measured strain	Northern truss	Measured strain	A (cm^2)	E (t/cm^2)	Force in northern truss (Tons)	Force in southern truss (Tons)	Force from numerical model (Tons)
SRS12	-184	SRN12	-169	64.3	2000	-22	-24	-23
SBS9	215	SBN9	184	140.3	2000	52	60	72
SUS8	-286	SUN8	-263	140.3	2000	-74	-80	-82
SBS7	299	SBN7	262	140.3	2000	73	84	102
SRS7	17	SRN7	22	64.3	2000	3	2	4
SRS6	51	SRN6	54	64.3	2000	7	7	6
SUS6	-362	SUN6	-359	140.3	2000	-101	-102	-104
SBS5	292	SBN5	238	140.3	2000	67	82	96
SUS4	-364	SUN4	-304	140.3	2000	-85	-102	-94
SBS3	209	SBN3	197	140.3	2000	55	59	76

The function 'TargetFunction2' defines a cost function for the model updating procedure.

1. A predefined force vector (force) is given, representing forces acting on different elements of the truss. A(21:27), a subset of the input parameter vector A, determines the distribution of forces between the southern truss (Sforce) and the northern truss (Nforce). Sforce gets the fraction of force determined by A(21:27), while Nforce gets the remaining fraction (1-A(21:27)).

2. Stiffness matrix k is applied to both Sforce and Nforce to compute the corresponding force-induced displacements (Skf and Nkf) in the south and north trusses, respectively.
3. 'Test' contains measured strain values from an experiment. The first column of test corresponds to the southern truss, and the second column to the northern truss. These strains are scaled by A(1:10) and A(11:20), respectively, representing correction factors for updating the model.
4. The function calculates the sum of squared differences between the scaled measured strains and the computed responses (Skf and Nkf). The objective (f) represents the error between the predicted and actual strain responses, and the optimization process aims to minimize this error.

Next, the following code is run to optimize the cost function:

```
x0=[64,140,140,140,64, 64,140,140,140,140,64,140,140,140,64, 64,140,140,1
40,140,.5,.5,.5,.5,.5,.5,.5];%area and load distribution, initial value
lb=[0, 0, 0, 0, 0, 0, 0, 0, 0, 0, 0, 0, 0, 0, 0,0,0, 0, 0, 0, .35, .35, .35, .35, .35, .35, .35];
%lower bound
ub=[91,150,150,150,91,91,150,150,150,150,91,150,150,150,91,91,150,150,150,
150,.65,.65,.65,.65,.65,.65,.65];];%upper bound of variables
[x,fval] = fmincon(@TargetFunction2,x0,[],[],[],[],lb,ub)
```

'x0' is the initial values for the optimization variables, which include:

- The areas of structural elements.
- The load distribution factors between the southern and northern trusses.

lb (lower bound) and ub (upper bound): These set constraints for the parameters, ensuring they remain within physically realistic ranges. 'fmincon' minimizes TargetFunction2 (which represents the error between measured and computed strain responses). It searches for an optimal set of x values that best fit the measured strain data while respecting the given bounds.

The output x contains the optimal values for the area parameters and load distribution factors that minimize the objective function.

The results are presented below. The optimized values for areas mostly converge near the upper limits (e.g., many values close to 150). The load distribution factors ($x(21{:}27)$) are spread across their bounds (0.35 to 0.65), with some staying at their limits. This suggests that the model favours higher material areas for structural stability and a relatively balanced load distribution.

```
x = [64.8972795722169   149.999999831565   149.999998640658   149.999999926575
     90.9999734818160   58.7562270754636   149.999999291785   149.999999902776
     136.306319253327   149.999999885153   63.5881688451855   149.999999823709
     148.032545276530   149.999999911516   90.9999664365274   53.8876450354951
     137.126432482908   149.999999909938   144.444075234043   149.999999858133
     0.649931576857285  0.350000056046724  0.649999934428646  0.649999906957287
     0.490384624294519  0.564730678026301  0.414551326737671]
```

6.5 NORMALIZATION OF DAMAGE-SENSITIVE FEATURES

Environmental and operational effects, such as humidity, wind, traffic, solar radiation, and temperature, significantly impact the defect damage index. These indices are crucial for accurately predicting the health of a structure. Therefore, monitoring environmental conditions in conjunction with the structural response is essential.

One of the key challenges in structural diagnosis is identifying the correlation between environmental and operational factors and their effects on defect-sensitive indices. This correlation is unique to each structure and necessitates long-term data collection. For instance, studies have shown that an increase in temperature typically leads to a reduction in the natural frequency of structures. This phenomenon can be attributed to thermal expansion, which alters the stiffness properties of the material. Conversely, a decrease in temperature below zero degrees Celsius may cause structural hardening due to the material's increased stiffness at lower temperatures [19].

Temperature also plays a decisive role in determining the boundary conditions of a structure. For example, thermal gradients can induce changes in boundary conditions, which may further complicate damage detection efforts. A comprehensive understanding of these environmental influences allows for better optimization of damage index, ultimately leading to more accurate damage diagnosis.

To quantify the impact of environmental factors on the damage index, it can be expressed as a function of multiple variables:

$$\mathrm{SI} = f(T, H, L)$$

where

- SI = Damage index
- T = Temperature
- H = Humidity
- L = Load (e.g., traffic load)

One way to quantify this relationship is through the following linear regression model:

$$\mathrm{SI}(T,\ H,\ L) = a_0 + a_1 T + a_2 H + a_3 L + \varepsilon$$

In this model:

- a_0, a_1, a_2, a_3 = Coefficients determined through regression analysis.
- ϵ = Error term, accounting for variability not explained by the model.

The coefficients provide insights into how each factor influences the damage index. For example, a negative a_1 coefficient indicates that an increase in temperature leads to a decrease in the damage index, suggesting a negative impact on defect detection capabilities.

To establish the correlation between environmental factors and damage index, Pearson correlation coefficient γ can be used, calculated as:

$$\gamma = \frac{N\sum XY - \sum X \sum Y}{\sqrt{\left(N\sum X^2 - \left(\sum X\right)^2\right)\left(N\sum Y^2 - \left(\sum Y\right)^2\right)}}$$

where

N = Number of observations
X = Independent variable (e.g., temperature)
Y = Dependent variable (e.g., index's sensitivity)

A coefficient value close to +1 indicates a strong positive correlation, while a value close to −1 indicates a strong negative correlation. Values near 0 imply no linear correlation, suggesting that the environmental factor does not significantly impact the damage index.

Long-term monitoring of structural responses, coupled with environmental conditions, provides a rich set of data that can be used to assess the performance and health of a structure over time. One of the tools that can be used to make sense of this data is time-series modelling. Specifically, the autoregressive integrated moving average (ARIMA) model is often applied to forecast future values of key indicators, such as the damage index, based on historical data.
The ARIMA model is represented as:

$$\text{ARIMA}(p,d,q)$$

where

- p is the order of the autoregressive (AR) part, which represents the influence of previous observations on the current value.
- d is the degree of differencing, which helps to make the data stationary (i.e., to remove trends and seasonality).
- q is the order of the moving average (MA) part, which models the influence of past forecast errors on the current value.

The ARIMA model is useful when we need to forecast future values based on patterns observed in past data. In structural health monitoring, for example, it can help forecast changes in the damage index, which is a measure of how sensitive a structure is to potential damage. By analysing past behaviour, the ARIMA model can predict future trends, helping engineers anticipate when defects may arise or when maintenance might be necessary.

To enhance the predictive capabilities of the damage index, advanced machine learning algorithms can be employed to model the relationships between environmental factors and damage index. Techniques such as regression analysis, support vector machines, and neural networks have been utilized in recent studies to refine these correlations and improve prediction accuracy [20].

For example, the optimization of the damage index can be framed as a mathematical optimization problem, where the objective function to be minimized could be the error between observed and predicted damage indices:

$$J(\theta) = \frac{1}{2}\sum_{i=1}^{N}\left(y_i - \widehat{y}_i\right)^2 + \lambda\sum_{j=1}^{m}\Theta_j^2$$

where

y_i = Actual output (e.g., observed damage index)
$\widehat{y}_i$ = Predicted output from the model
λ = Regularization parameter
θ = Model parameters

The optimization process typically employs gradient descent to minimize the objective function iteratively as shown below (details are as explained in Section 6.4.4)

$$\theta^{(t+1)} = \theta^{(t)} - \alpha\nabla J\left(\theta^{(t)}\right)$$

This iterative process continues until convergence is achieved, ensuring that the damage index accurately reflects the structural condition.

Figure 6.11 illustrates the variations in the first natural frequency in vertical and lateral directions over a three-month period, which were extracted at a specific time in a day. The seasonality and ambient temperature cause the natural frequencies to change.

By integrating these mathematical formulations and optimization techniques into the evaluation of damage indices, engineers can refine their predictions and enhance the reliability of structural health monitoring. These approaches allow for a more nuanced understanding of how environmental and operational factors influence the structural response, ultimately leading to more accurate and timely damage diagnoses. The combination of machine learning and robust statistical methods equips engineers with the necessary tools to navigate the complexities of structural health monitoring in a variable environment.

6.6 DAMAGE DETECTION TECHNIQUES

After determining the damage-sensitive index and extracting it from the measured signal—along with removing the effects of environmental and operational factors—damage detection method is applied to distinguish between normal and damaged states. As mentioned in Table 6.1, if the damage-sensitive indices are known in different damage states, classification techniques are used, and if the damage-sensitive indices are not known, outlier analysis should be used.

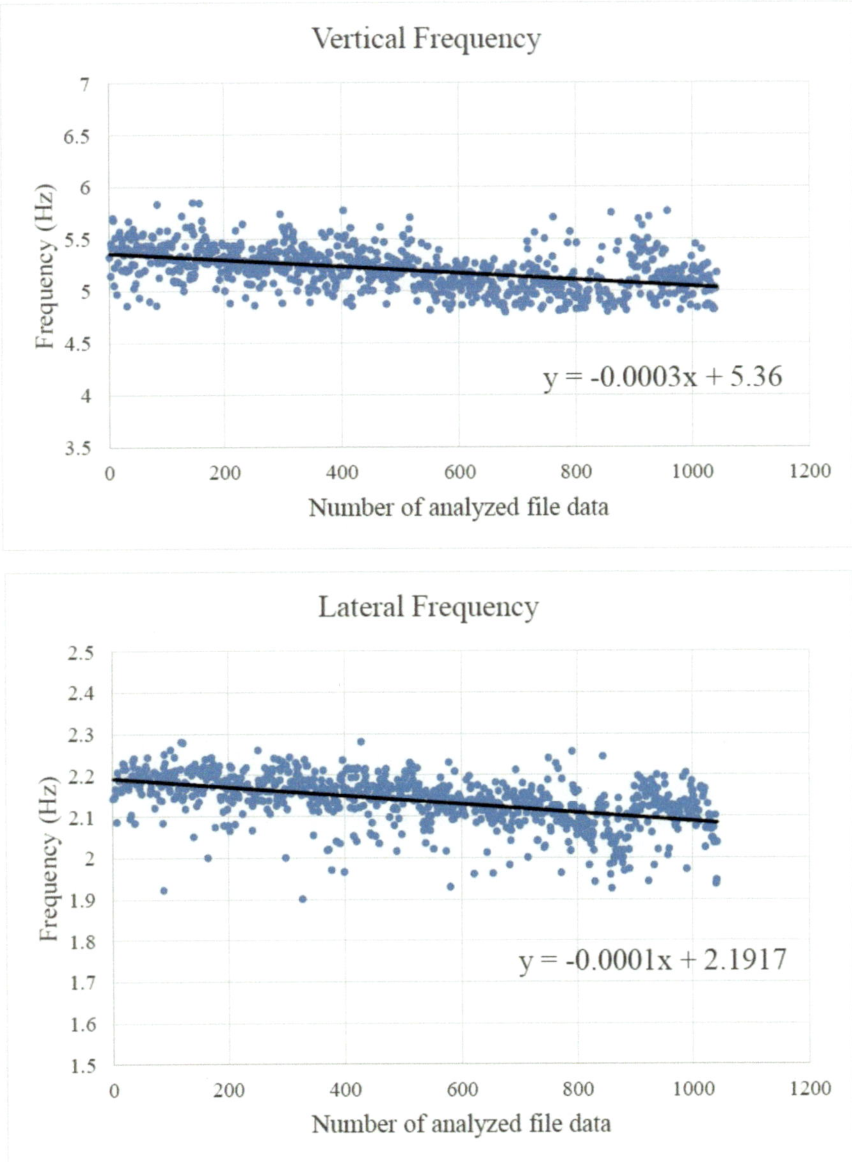

FIGURE 6.11 Pattern of change of natural frequency over a three-month period.

6.6.1 Classification Methods

Some of the common classification methods include hypothesis testing, hidden Markov models (HMMs), and 6.7.1.3. Multi-layer feed forward neural networks (MLP NNs).

6.6.1.1 Hypothesis Testing

Hypothesis testing is a statistical approach that examines whether observed data differ significantly from a predefined null hypothesis, suggesting potential damage if the null hypothesis is rejected. The null hypothesis, H_0, typically assumes no damage, and the test statistic T is calculated from observed data. When T exceeds a threshold (based on the chosen significance level), H_0 is rejected, indicating potential damage:

$$T = \frac{\bar{X} - \mu_0}{\frac{\sigma}{\sqrt{n}}}$$

where $\bar{X}$ is the sample mean, μ_0 is the hypothesized mean (representing the normal state), σ is the standard deviation, and n is the sample size.

For instance, Goi [21] conducted a study on a bridge that experienced frequent heavy loads, leading to suspected changes in its structural integrity. By analysing the bridge's vibration data with hypothesis testing, they were able to detect variations in the vibration intensity that indicated damage. This method was effective because hypothesis testing can provide a quantitative means to assess whether the observed changes in the structural response are statistically significant, minimizing the influence of random fluctuations.

6.6.1.2 Hidden Markov Models (HMMs)

Hidden Markov models (HMMs) are particularly effective for analysing systems with hidden states, where the structure's behaviour changes over time in response to evolving operational loads. In an HMM, hidden states are represented by $Q=\{q_1, q_2, \ldots, q_N\}$, and observable variables by $O=\{o_1, o_2, \ldots, o_T\}$, governed by transition probabilities $P(q_j|q_i)$ indicating the likelihood of transitioning from one hidden state to another.

To determine the optimal HMM parameters, the Baum–Welch algorithm is used, an iterative method for estimating parameters that maximize the probability of observing a given sequence. The algorithm adjusts parameters to maximize the probability of observing the sequence O with the HMM model parameters λ. The probability $P(O\lambda)$ is computed by summing over all possible hidden state sequences Q:

$$P(O\lambda) = \sum_{Q} P(OQ,\lambda) P(Q\lambda)$$

where λ denotes the model parameters.

Li et al. [22] applied an improved HMM to monitor railway infrastructure by analysing fibre optic sensor signals. Their approach effectively captured the temporal evolution of the track bed's structural response under varying load conditions. By interpreting changes in the signal patterns over time, the HMM was able to detect and classify defects, such as track bed deterioration, with high accuracy. This demonstrated the model's capability to distinguish between normal and abnormal structural states, making it a valuable tool for predictive maintenance in railway systems.

6.6.1.3 Multilayer Feed Forward Neural Networks (MLP NNs)

MLP networks are versatile models inspired by the human neural structure, commonly used in damage detection for their ability to learn complex patterns. Figure 6.12 shows a real neuron and its mathematical representation. Artificial neural networks

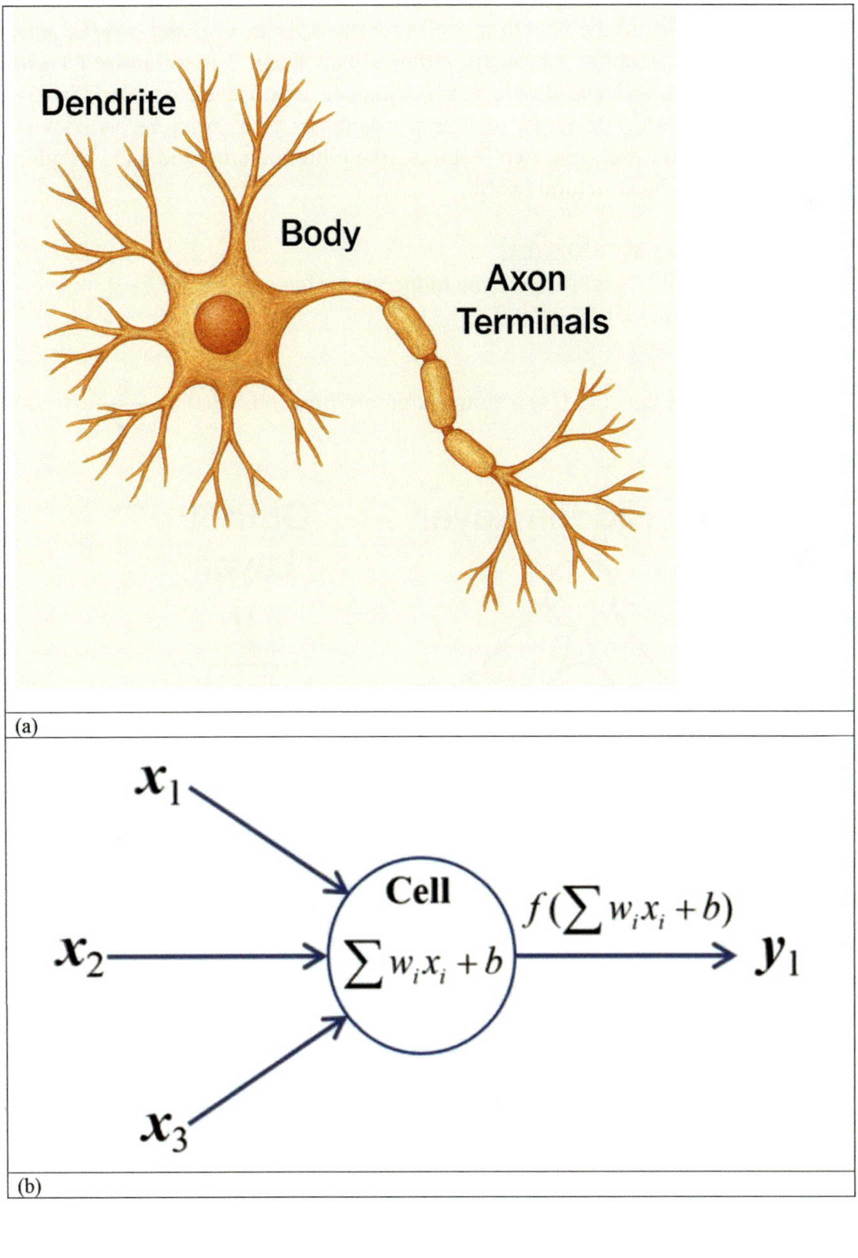

FIGURE 6.12 (a) A human neuron and (b) its mathematical representation [23].

(ANNs) learn through iterative adjustments of weights and biases to minimize prediction errors.

The forward propagation of data through the network can be represented as:

$$a^{(l+1)} = \sigma\left(W^{(l)}a^{(l)} + b^{(l)}\right)$$

where $W^{(l)}$ and $b^{(l)}$ are the weight and bias matrices for layer l, and σ is the activation function. The backpropagation algorithm adjusts W and b to minimize the error.

For example, consider a simple ANN with one input layer, one hidden layer, and one output neuron designed to identify damages in a structure, as shown in Figure 6.13. Suppose we input two features, like vibration amplitude and frequency shift, representing the structural health.

1. Input vector: $a^{(0)} = [0.5, 0.3]$
2. Weights and biases: Assume the hidden layer has weights $W^{(1)} = \begin{bmatrix} 0.2 & 0.4 \\ 0.1 & 0.3 \end{bmatrix}$ and biases $b^{(1)} = [0.1, 0.2]$.
3. Activation Function: Use a simple sigmoid function, $\sigma(z) = \dfrac{1}{1+e^{-z}}$

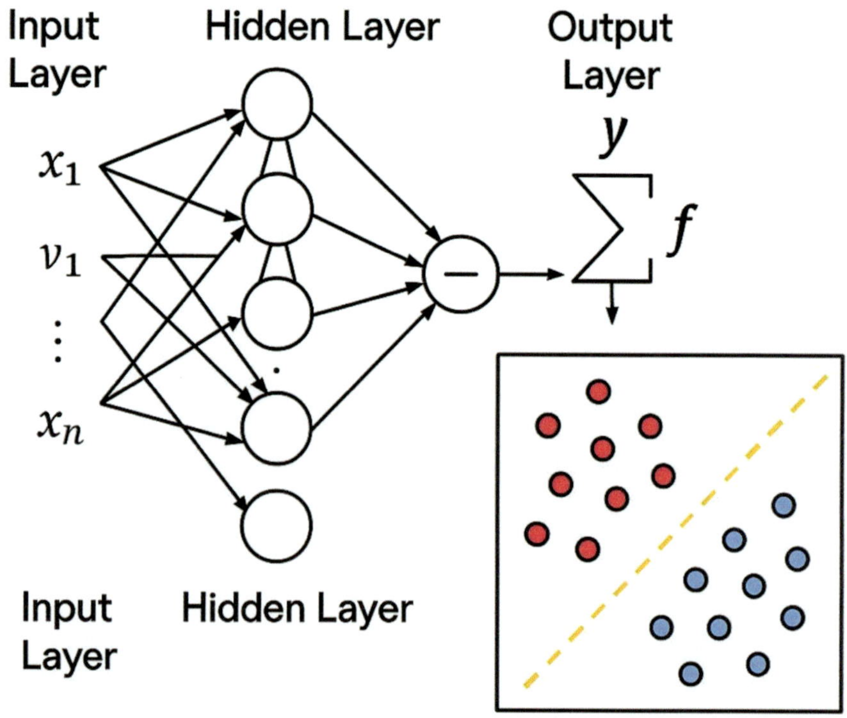

FIGURE 6.13 Simple ANN with one input layer, one hidden layer, and one output neuron.

For the first hidden neuron:

$$z_1 = (0.2 \times 0.5) + (0.4 \times 0.3) + 0.1 = 0.36$$

Applying the activation function:

$$a_1^{(1)} = \sigma(0.36) \approx 0.59$$

Repeating for all neurons, the output layer neuron provides a final value indicating a potential damage (e.g., close to 1 for a high likelihood of damage).

Backpropagation then computes the error and adjusts W and b to improve predictions over many iterations. This process enables the ANN to accurately learn patterns associated with damages, making it effective for structural monitoring and diagnosis. MLP has hidden layers (multi-layer perception) and could solve nonlinear classification problems as well.

Lu et al [24] and Ataei et al [25] explored the use of statistical modelling to predict deterioration in bridge components, where accurate forecasting is essential for optimizing maintenance and preventing unexpected failures. They employed an ordinal logistic regression model trained on historical inspection data to estimate condition ratings for elements such as bridge bearings, expansion joints, and girders. The model effectively identified degradation trends and provided actionable insights for prioritizing maintenance activities. This approach demonstrated the value of data-driven prediction methods in environments where manual inspection may be limited or delayed.

6.6.1.4 Convolutional Neural Networks (CNNs)

CNNs are designed to process grid-like data, such as time-series vibration data transformed into spectrograms. A spectrogram is a visual representation of the spectrum of frequencies in a signal as it varies with time. CNNs apply convolutional filters to detect localized features, making them well-suited for detecting local damages.

In a landmark study, Jiang, Pang, and Li [26] developed a CNN-based model for the rapid detection and classification of damage in concrete structures. The researchers utilized image data to train the model, enabling it to distinguish between various damage types such as cracks and spalling. Their approach achieved high classification accuracy, demonstrating the effectiveness of CNNs in automating damage assessment. This study underscored the potential of deep learning to enhance inspection efficiency, particularly in large and complex infrastructure systems.

6.6.1.5 Recurrent Neural Networks (RNNs)

Recurrent neural networks (RNNs) are particularly effective in processing sequential data because they retain information across time steps. This makes them suitable for applications like monitoring structures where damage progression occurs over time. Long short-term memory (LSTM) networks are a specialized type of RNN designed to overcome the vanishing gradient problem, allowing them to capture long-range dependencies in sequential data.

The hidden state in LSTMs, updated at each time step t, is given by:

$$h_t = f\left(h_{t-1}, x_t\right)$$

where f represents the function that controls the flow of information through the LSTM cells. This function typically involves gates that regulate the forgetting, updating, and retention of information, allowing LSTMs to effectively capture long-term dependencies in data sequences. LSTMs are particularly valuable in SHM, where damage may evolve gradually, and capturing these long-term temporal patterns is crucial.

For example, consider a scenario where vibration data is collected over time from a bridge. Each data point x_t represents the vibration measurement at time step t. The goal is to use an LSTM to predict the future structural state and detect potential damage progression over time.

At time $t-1$, the LSTM updates its hidden state h_{t-1}, which contains information from the previous time steps. At time t, a new vibration measurement x_t is input into the LSTM, and the hidden state h_t is updated based on both the previous hidden state h_{t-1} and the new input x_t. This updated state captures both the short-term and long-term changes in the structure's behaviour.

For example, if the vibration data x_t suddenly deviates from the normal pattern, the LSTM will adjust its internal state and may predict an increasing likelihood of damage. Over time, as the LSTM continues to process new data, it learns to recognize patterns that signal the onset or progression of structural issues.

Zhang et al. [27] proposed a novel method for predicting crack propagation in concrete gravity dams using a LSTM neural network. The researchers generated simulated crack data through the extended finite element method (XFEM) and used this data to train the LSTM model. Their approach enabled accurate forecasting of crack growth paths, demonstrating the model's ability to capture the temporal evolution of structural damage. This study highlighted the potential of LSTM networks in predictive maintenance scenarios, particularly when high-quality simulation data is available for training.

6.6.2 Outlier Analysis Techniques

Outlier analysis techniques are valuable in situations where only normal-state data is available. These methods learn the normal behaviour of a structure, flagging deviations as potential damages. These deviations are identified based on their distance, or relative density from other points. K-means and statistical methods are based on distance, whereas DBScan is based on density.

6.6.2.1 Statistics Methods

Statistical methods are among the most traditional approaches for outlier detection. These methods rely on the assumption that the data follows a specific statistical distribution, such as a normal (Gaussian) distribution. Outliers are identified as data points that deviate significantly from the expected range based on this distribution.

For example, one common statistical method is the *Z*-score technique. In this approach, the mean and standard deviation of the dataset are calculated, and each data point is assigned a *Z*-score, which measures how far it lies from the mean in terms of standard deviations. Points with *Z*-scores exceeding a certain threshold (e.g., ±3) are flagged as outliers. Another popular method is the interquartile range (IQR) technique, which identifies outliers as points that fall outside the range defined by 1.5 times the IQR below the first quartile (Q1) or above the third quartile (Q3).

The main advantage of statistical methods is their simplicity and efficiency. They are easy to implement and work well for small datasets with clear distribution patterns. However, these methods have limitations. They often assume that the data follows a specific distribution (e.g., normal), which may not always be true in real-world scenarios. Additionally, they struggle with high-dimensional data or datasets with complex structures.

Statistical methods are best suited for univariate or low-dimensional datasets where the underlying data distribution is well understood and relatively simple.

6.6.2.2 K-Means Method

The K-means method is a clustering algorithm that can also be used for outlier detection by identifying points that do not fit well into any cluster. The algorithm works by partitioning the dataset into a predefined number of clusters (denoted as k). It starts by randomly selecting k initial cluster centroids and then iteratively assigns each data point to the nearest centroid based on distance (usually Euclidean). After all points are assigned, the centroids are recalculated as the mean of all points in their respective clusters. This process repeats until the centroids stabilize.

Once clusters are formed, outliers can be identified as points that lie farthest from their assigned cluster centroid. For instance, a threshold can be set based on a certain percentile of distances, and points exceeding this threshold are flagged as outliers.

The K-means method has several advantages. It is computationally efficient and works well for large datasets with compact, spherical clusters of similar sizes. However, it also has notable drawbacks. The algorithm requires predefining the number of clusters (k), which may not always be known in advance. It is also sensitive to the initial placement of centroids, which can lead to suboptimal clustering results if not chosen carefully. Furthermore, K-means struggles with datasets containing non-spherical or overlapping clusters.

This method is most effective when dealing with datasets that have distinct and well-separated clusters and when computational efficiency is a priority [25].

6.6.2.3 DBSCAN Method

Density-based spatial clustering of applications with noise (DBSCAN) is a powerful density-based clustering algorithm that excels at identifying outliers in datasets with varying densities or irregularly shaped clusters. Unlike K-means, DBSCAN does not require specifying the number of clusters beforehand; instead, it relies on two parameters: 'ε' (epsilon), which defines the radius of a neighbourhood around each point, and 'minPts', which specifies the minimum number of points required to form a dense region.

The algorithm works by classifying each point in the dataset into one of three categories: core points, border points, or noise (outliers). Core points are those that have at least minPts neighbours within their ε-radius neighbourhood. Border points are those that fall within the ε-radius of a core point but do not themselves have enough neighbours to qualify as core points. Noise points are those that do not belong to any cluster and are considered outliers.

Clusters are formed by expanding from core points: any point within an ε-radius neighbourhood of a core point is added to its cluster, and this process continues recursively until no more points can be added.

DBSCAN has several strengths that make it particularly useful for outlier detection. It can handle datasets with arbitrary-shaped clusters and varying densities, making it more flexible than K-means in many scenarios. Additionally, it automatically identifies noise points (outliers) without requiring separate post-processing steps.

However, DBSCAN has its limitations as well. Its performance is highly sensitive to the choice of parameters (ε and minPts), which can be challenging to tune effectively for complex datasets. It may also struggle with datasets containing clusters of widely differing densities or very high-dimensional data.

DBSCAN is best used when working with datasets that have irregularly shaped clusters or when outliers reside in sparse regions that do not belong to any dense cluster.

6.7 SUMMARY

This chapter provided a comprehensive exploration of damage diagnosis techniques in structural health monitoring. The chapter began with an introduction to the concept of damage diagnosis and its relevance to ensuring the structural integrity of critical infrastructures. It then defined the concept of a damage in engineering contexts, discussing common causes of damage to structures, including environmental and operational factors that contribute to wear and deterioration.

Building on these definitions, the chapter introduced various damage-sensitive indicators that serve as measurable features capable of detecting and characterizing damage within structures. These indicators were systematically categorized into scalar, vector, and image-based types, each offering unique insights depending on the nature of the data and the damage being assessed. Scalar indicators, such as natural frequencies and modal damping ratios, provide simplified numerical values that reflect changes in structural behaviour. Vector indicators extend this by incorporating directional information, capturing more complex variations in structural responses. Image-based indicators leverage advanced visual data processing techniques, including thermal imaging and digital image correlation, to localize and quantify damage with high spatial resolution.

Additionally, the chapter presented a numerical model validation algorithm designed to update and calibrate finite element models based on observed data, enhancing the accuracy of damage localization and severity estimation. This was exemplified through a detailed case study involving the Neka bridge, where numerical model updating techniques were applied to real-world data, demonstrating the practical utility of the methods discussed.

Normalization of damage-sensitive features was also addressed, emphasizing the importance of preprocessing data to reduce variability caused by environmental and operational changes. This step ensures that damage detection techniques focus on true structural anomalies rather than noise or external influences.

The chapter further explored a range of damage detection techniques, highlighting classical statistical methods alongside modern machine learning and outlier analysis approaches. These techniques enable the identification of subtle changes in structural behaviour that precede visible damage, facilitating early intervention and maintenance planning.

Together, these sections presented a thorough overview of damage diagnosis, combining traditional methods with state-of-the-art machine learning approaches to equip engineers and researchers with the tools needed to monitor, predict, and mitigate structural failures.

REFERENCES

[1] C. R. Farrar and K. Worden, *Structural Health Monitoring: A Machine Learning Perspective.* Wiley, 2013.

[2] M. Qiao, D. K. Kim, J. H. Yi, and Y. Chen, "Baseline Models for Bridge Performance Monitoring," *Journal of Engineering Mechanics*, vol. 130, no. 5, pp. 562–569, 2004.

[3] G. Chen, I. Alomari, W. Taffese, and M. H. Afsharmovahed, "Earthquake-induced Damage Classification of Bridges Using Artificial Neural Network," Mid-America Transportation Center, pp. 1–45, 2024.

[4] Y. Wang, X. Zhang, L. Liu, and H. Li, "Comprehensive Review of the Effects of Vibrations on Wind Turbine During Energy Generation Operation," in *Proceedings of the International Conference on Intelligent Manufacturing and Industrial Automation*, Springer, pp. 887–896, 2020.

[5] Y. O. Busari, Y. H. P. Manurung, W. E. W. A. Rahaman, et al., "Fatigue Crack Growth Analysis of Welded Structural Steel S460G2+M," *International Journal of Advanced Manufacturing Technology*, vol. 126, pp. 1235–1248, 2025.

[6] S. Miao, L. Su, C. Zhang, et al., "Dynamic Response Characteristics and Damage Failure Process of Bedding Rock Slope," *Bulletin of Engineering Geology and the Environment*, vol. 83, no. 2, pp. 1–15, 2024.

[7] R. Soleimanpour and C.-T. Ng, "Scattering Analysis of Nonlinear Lamb Waves at Delaminations in Composite Laminates," *Ultrasonics*, vol. 111, pp. 106309, 2021.

[8] Z. Yi, J. Ma, and L. Gao, "The Silver Bridge Collapse: Welding Fatigue Defects Cause Trouble," *Metal World*, vol. 2023, no. 5, pp. 14–20, 2023.

[9] P. Paultre, É. Calais, J. Proulx, C. Prépetit, and S. Ambroise, "Damage to Engineered Structures During the 12 January 2010, Haiti Earthquake," *Canadian Journal of Civil Engineering*, vol. 38, no. 7, pp. 757–767, 2011.

[10] Y.-R. Dong, Z.-D. Xu, Y.-S. Xu, S. Waseem, and Q.-Q. Li, "Seismic Performance and Multi-scale Model of Damage Evolution for Reinforced Concrete Frame Structures," *ACI Structural Journal*, vol. 118, no. 2, pp. 3–15, 2021.

[11] Y. Lan, W. Lin, and Y. Zhang, "Bridge Frequency Identification Using Vibration Responses from Sensors on a Passing Vehicle," in *Bridge Safety, Maintenance, Management, Life-Cycle, Resilience and Sustainability*, CRC Press, pp. 1–8, 2022.

[12] D. P. Granados, M. A. Ortega Ruiz, J. M. Acosta, S. A. Gama Lara, R. A. González Domínguez, and P. J. Páramo Kañetas, "A Wind Turbine Vibration Monitoring System for Predictive Maintenance Based on Machine Learning Methods Developed Under Safely Controlled Laboratory Conditions," *Energies*, vol. 16, no. 5, pp. 2290, 2023.

[13] A. Cancelli, F. Ubertini, S. Laflamme, A. Alipour, and S. Sritharan, "Vibration-based Damage Localization and Quantification in a Pretensioned Concrete Girder Using Stochastic Subspace Identification and Particle Swarm Model Updating," *Structural Health Monitoring*, vol. 19, no. 2, pp. 587–605, 2020.

[14] A.-T. Aman, C. Tufisi, G.-R. Gillich, and Z. I. Praisach, "Detection of Transverse Cracks in Steel Beams Using Damage Location Coefficients and Artificial Neural Networks," *Journal of Vibration Engineering & Technology*, vol. 11, no. 3, pp. 345–359, 2020.

[15] R. Shafique, H.-U.-R. Siddiqui, F. Rustam, S. Ullah, M. A. Siddique, E. Lee, I. Ashraf, and S. Dudley, "A Novel Approach to Railway Track Faults Detection Using Acoustic Analysis," *Sensors*, vol. 21, no. 18, pp. 6221, 2021.

[16] J. Chakraborty, X. Wang, and M. Stolinski, "Damage Detection in Multiple RC Structures Based on Embedded Ultrasonic Sensors and Wavelet Transform," *Buildings*, vol. 11, no. 2, pp. 56, 2021.

[17] S. Setsobhonkul, S. Kaewunruen, and J. M. Sussman, "Lifecycle Assessments of Railway Bridge Transitions Exposed to Extreme Climate Events," *Frontiers in Built Environment*, vol. 3, pp. 1–15, 2017.

[18] F. Xiao, J. L. Hulsey, G. S. Chen, and Y. Xiang, "Optimal Static Strain Sensor Placement for Truss Bridges," *International Journal of Distributed Sensor Networks*, vol. 13, no. 5, pp. 1–13, 2017.

[19] A. Poudel, S. Kim, B. H. Cho, and J. Kim, "Temperature Effects on the Natural Frequencies of Composite Girders," *Applied Science*s, vol. 14, no. 3, pp. 1175, 2024.

[20] M. Flah, I. Nunez, W. Ben Chaabene, and M. L. Nehdi, "Machine Learning Algorithms in Civil Structural Health Monitoring: A Systematic Review," *Archives of Computational Methods in Engineering*, vol. 28, no. 6, pp. 2621–2643, 2021.

[21] Y. Goi, "Bayesian Damage Detection for Vibration-based Bridge Health Monitoring," Ph.D. dissertation, Department Civil and Earth Resources Engineering, Kyoto University, Kyoto, Japan, 2018.

[22] W. Li, L. He, Z. Li, and Y. Wan, "Detection of Track Bed Defects Based on Fibre Optic Sensor Signals and an Improved Hidden Markov Model," *Electronics*, vol. 13, no. 13, pp. 2504, 2024.

[23] W. S. McCulloch and W. Pitts, "A Logical Calculus of the Ideas Immanent in Nervous Activity," *Bulletin of Mathematical Biophysics*, vol. 5, pp. 115–133, 1943.

[24] P. Lu, H. Wang, and D. Tolliver, "Prediction of Bridge Component Ratings Using Ordinal Logistic Regression Model," *Mathematical Problems in Engeering*, vol. 2019, Article ID 9797584, pp. 1–11, 2019.

[25] S. Ataei, A. A. Aghakouchak, M. S. Marefat, and S. Mohammadzadeh, "Sensor Fusion of a Railway Bridge Load Test Using Neural Networks," *Expert System with Applications*, vol. 29, no. 3, pp. 678–683, 2005.

[26] Y. Jiang, D. Pang, and C. Li, "A Deep Learning Approach for Fast Detection and Classification of Concrete Damage," *Automation in Construction*, vol. 122, pp. 103785, 2021.

[27] K. Zhang, F. Lu, Y. Peng, and X. Li, "A Novel Method for Generation and Prediction of Crack Propagation in Gravity Dams," *Structural Engineering and Mechanics*, vol. 81, no. 6, pp. 665–675, 2022.

7 Infrastructure Service Life Planning

7.1 INTRODUCTION

Design life refers to the intended lifespan of a structure or component, while service life planning involves strategies to ensure the infrastructure meets or exceeds its intended lifespan. This process includes assessing design parameters, materials, environmental exposure, and expected loads to estimate deterioration rates and maintenance needs.

Several factors influence an asset's service life, including material properties, environmental conditions, loading patterns, and maintenance history. For example, railway tracks subjected to high-frequency dynamic loading degrade faster than those experiencing minimal load variations. Similarly, coastal bridges deteriorate more rapidly due to saltwater exposure and humidity, accelerating corrosion.

The ISO 55000 framework emphasizes a lifecycle approach to asset management, ensuring that infrastructure assets deliver value throughout their lifespan. This aligns with the Maintenance Management Excellence Model (MMEM), which applies ISO 55000 principles to maintenance by implementing structured, data-driven approaches. The final goal is to adopt maintenance measures that can extend the life of the asset. Proper maintenance of an asset therefore requires effective service life planning.

7.2 SERVICE LIFE PLANNING

The schematic degradation model of a structural component or section resembles the curve shown in Figures 7.1 and 7.2. In such models, the degradation rate is initially low but accelerates significantly over time. This behaviour is sometimes referred to as the 'bathtub curve' in reliability engineering, where the initial phase represents the 'infant mortality' period, followed by a steady-state phase, and finally, the wear-out phase where degradation accelerates [1]. Service life planning should be structured to ensure that repair actions are performed before the degradation rate becomes too rapid, as this can lead to catastrophic failures and increased costs [2].

The prediction of the degradation rate can be based on manufacturer recommendations, historical experience, or structural monitoring data. Furthermore, knowing the reference service life, an estimate of the remaining service life can be

DOI: 10.1201/9781003494652-7

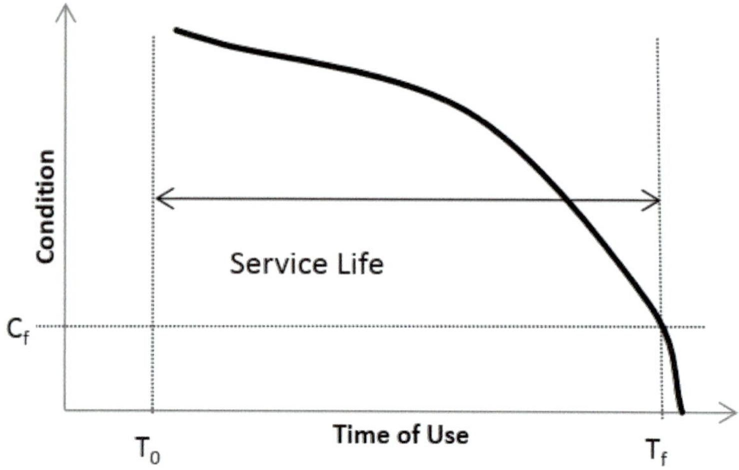

FIGURE 7.1 Structure degradation model.

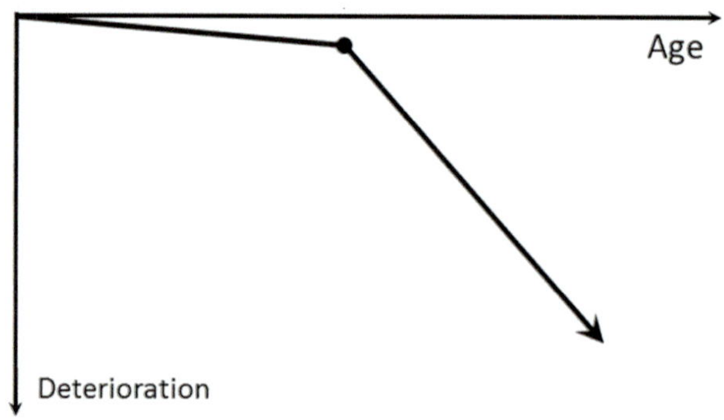

FIGURE 7.2 Simplified degradation model of a structure.

achieved using coefficient-based methods. These methods account for factors such as construction quality, environmental conditions, and operational usage [3].

The coefficient method for estimating service life involves multiplying the reference service life (RSLC) by a series of corrective coefficients [3]. These coefficients adjust for various influencing factors and can take values of 0.8, 1, or 1.2. The formula is expressed as

$$\text{ESLC} = \text{RSLC} \times F_A \times F_B \times F_C \times F_D \times F_E \times F_F \times F_G$$

where

- ESLC is the estimated service life
- RSLC is the reference service life

- F_A the component quality (1.2, 1, 0.8)
- F_B the design level (1.2, 1, 0.8)
- F_C the construction quality (1.2, 1, 0.8)
- F_D the internal environmental conditions (1.2, 1, 0.8)
- F_E the external environmental conditions (1.2, 1, 0.8)
- F_F the operational conditions (1.2, 1, 0.8)
- F_G the maintenance level (1.2, 1, 0.8)

For fatigue life prediction, the AASHTO code can be applied, while corrosion life prediction can utilize models proposed by researchers. An example of fatigue life prediction is presented in Chapter 4.

Integrating these methods into service life planning ensures that repair actions are performed before the degradation rate becomes too rapid, as this can lead to catastrophic failures and increased costs [2], therefore extending the service life of the structure and mitigating costly repairs.

7.3 MAINTENANCE EXCELLENCE MODEL AND SERVICE LIFE PLANNING

Step 5 of the maintenance excellence model presented in Chapter 1 provides a systematic approach to asset life planning through the following concepts, as shown in Figure 7.3:

- Life cycle cost analysis (LCCA)
- Predictive maintenance strategies
- Incorporating SHM as part of asset life cycle management

Life cycle cost (LCC) analysis is a critical decision-making tool that considers all costs associated with an infrastructure asset throughout its entire lifespan. These costs include initial construction, maintenance, operational expenses, repairs, and eventual decommissioning. Unlike traditional budgeting approaches that focus primarily on upfront costs, LCC analysis provides a more holistic view by integrating long-term financial planning.

By conducting LCCA, asset managers and engineers can

- evaluate the long-term economic viability of various design and maintenance strategies;
- compare multiple alternatives based on their total cost over time rather than initial cost alone;
- optimize maintenance and rehabilitation schedules to minimize overall expenditures; and
- justify investments in advanced monitoring and maintenance technologies by quantifying their cost benefits.

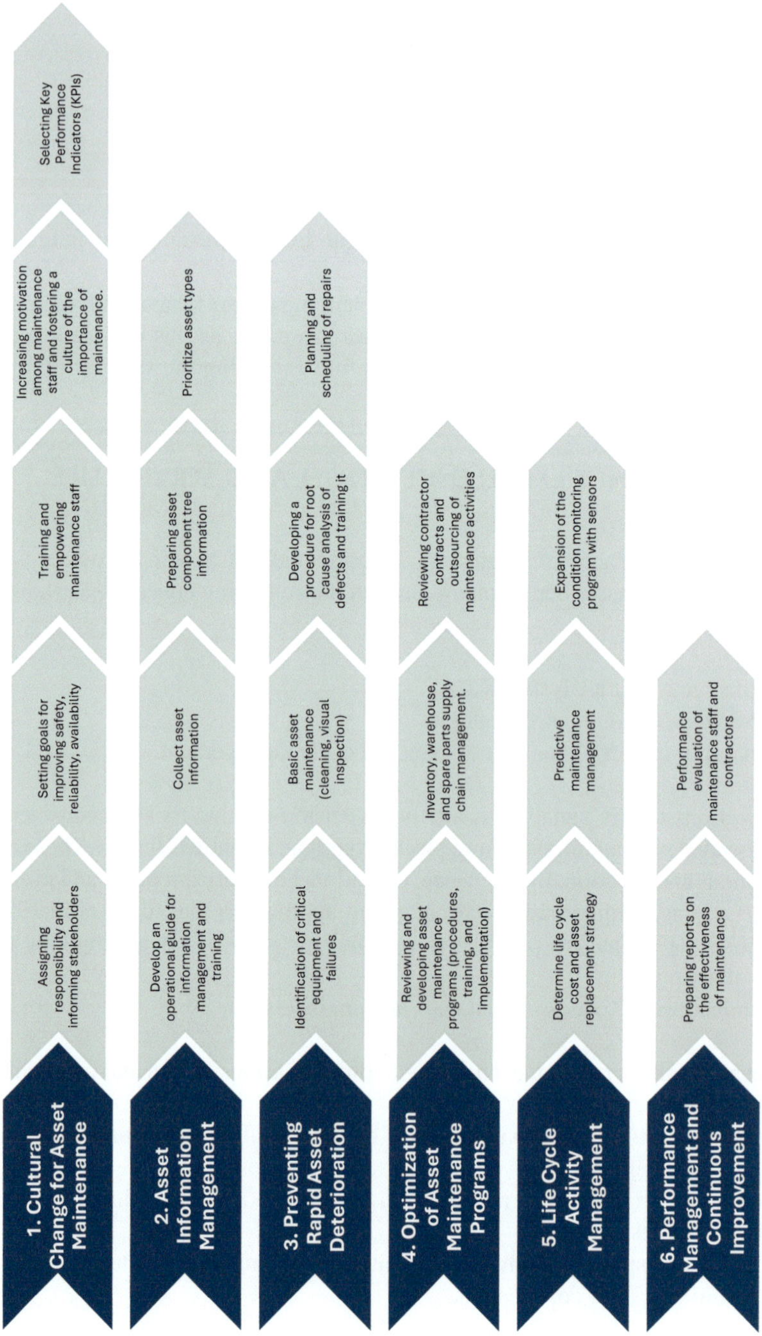

FIGURE 7.3 Simplified maintenance excellence model.

For example, a high-quality material with a higher initial investment may have lower overall costs due to reduced maintenance and a longer service life.

Predictive maintenance strategies focus on proactively addressing potential infrastructure issues before they escalate into critical failures. Traditional reactive maintenance approaches, which involve repairing assets only after a failure has occurred, often lead to costly repairs, extended downtimes, and compromised safety. In contrast, predictive maintenance leverages real-time data and analytical models to anticipate deterioration trends and schedule interventions at the optimal time.
Key benefits of predictive maintenance include:

- *Prolonging asset lifespan*: Addressing minor issues before they escalate significantly extends service life.
- *Reducing maintenance costs*: Early detection of defects reduces the need for expensive emergency repairs.
- *Enhancing operational efficiency*: Predictive analytics improve resource allocation and minimize unnecessary maintenance interventions.
- *Improving safety and reliability*: Regular assessments ensure that infrastructure remains in optimal working condition, mitigating risks associated with sudden failures.

By integrating predictive maintenance within MEM, infrastructure managers can shift from reactive to proactive strategies, ensuring that assets remain functional and cost-effective over their entire service life.

It is evident that SHM can significantly improve the service life planning of the structure and assist in improving the LCC and adopting predictive maintenance strategies for assets. SHM systems can provide accurate, real-time data on the condition of infrastructure components. By analysing trends in structural performance data, engineers can predict when components are likely to fail and schedule maintenance accordingly, optimizing resource allocation and minimizing downtime.

For example, SHM can track track-bed deformations, rail wear, and loading conditions. Using these data, it is possible to generate track condition deterioration models. Consequently, instead of interval-based resurfacing of the track, deterioration models can be used to create maintenance schedules that will reduce overall costs and operational disruptions.

Additionally, SHM systems will assist in having a better understanding of both capacity and demand on structures. These can include a better understanding of the hidden capacity of the asset components, as well as identifying more accurate operational loads on assets. By incorporating these, it is possible to extend the service lifetime or adjust maintenance intervals through more refined and accurate assessments.

7.4 APPLICATION OF SHM IN ASSET LIFE CYCLE MANAGEMENT

In general, structural capacity must exceed the demand induced by code-specified loads. However, in practice, this relationship may not hold for certain constructed structures. In such cases, two options are available:

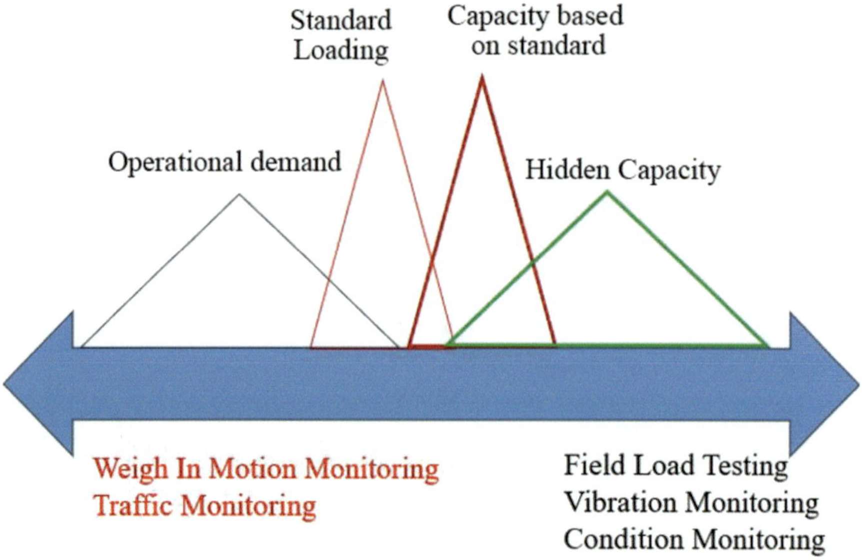

FIGURE 7.4 Role of structural monitoring and traffic monitoring in asset maintenance decision-making.

1. Operational Restrictions: Imposing limits on usage.
2. Structural Strengthening: Reinforcing the structure.

In many instances, however, structural elements may possess hidden capacity, or the actual loads may be less than the code-specified loads. Structural monitoring can determine the extent of hidden capacity or actual load conditions, as shown in Figure 7.4. Within this framework, numerous structures theoretically requiring restrictions or strengthening might, in reality, not require such measures.

This chapter will focus on how SHM systems can assist in determining hidden capacity and actual load conditions. Several case studies on bridge monitoring, rail monitoring, and traffic monitoring (e.g., moving axle loads) will be discussed to illustrate the application of structural monitoring in this context.

7.5 IDENTIFICATION OF HIDDEN CAPACITY OF STRUCTURES

As described in the previous section, civil structures may have hidden capacity that are not covered by standards, and therefore, will be able to withstand additional loads. Understanding the magnitude of hidden capacity is a rather difficult task, yet through implementing SHM, it is possible to get a better understanding of it. In the following sections, two SHM systems are described, which are installed on a bridge and a turnout structure in the Iranian railways. The goal of these long-term monitoring systems was to understand the response of the structures to passing loads and get a better understanding of their operational characteristics, rather than design values. These operational characteristics can then assist in making decisions regarding increasing the axle load or operational speed on these structures.

7.5.1 Veresk Bridge Monitoring System

Veresk bridge is one of the oldest arch bridges in Iran. With a total length of 75 m, it is the longest single span masonry arch bridge in Iran. The bridge accommodates a single ballasted track, as shown in Figure 7.5.

Initially, the route was designed for an axle load of 20 tonnes. Currently the operational axle load of the bridge is 20 tonnes [5]. Given the age of the structure, asset

FIGURE 7.5 (a) A side view of Veresk masonry arch bridge and (b) dimension of the bridge and location of strain gauges (red circles).

manager requested an assessment of this bridge to ascertain its continuous safe performance. In answering this question, several issues were raised:

- What is the current deterioration level of the material?
- What dynamic amplification factor should be considered for the bridge load assessment?
- What modules of elasticity will best represent the combined masonry block and mortar material?

Although Iranian design code provided values for the above parameters, these were very conservative and would render the bridge unsafe. However, the bridge had operated for more than 75 years without any issues. Therefore, a decision was made to design a long-term monitoring system to be deployed on the bridge to provide a comprehensive overlook of the response of the structure, as well as the type of traffic crossing over it.

This system comprises strain gauges that were strategically installed at various locations on the bridge, as shown in Figure 7.5b. These strain gauges were installed to measure the strain in between masonry blocks. Another set of strain gauges were installed on rails, which served as weigh in motion (WIM) sensors (Figure 7.6). Also, two sets of axle counters were installed at each end of the bridge to both activate the system and measure the number of axles on each passing train.

Using a set of electronic boards (Figure 7.7), a data collection unit has been created that can collect data from all channels. Strain gauges, displacement sensors, and vibration sensors are connected to this unit. A high-processing-power computer receives the signals and stores them after filtering. In this system, the health status and traffic of the structure can be observed and tracked at any moment.

Once the sensors were installed on the bridge, their functionality was verified through repeated measurements. If the sensors are properly installed and functioning correctly, the passage of the same train over the bridge should produce consistent results. This is demonstrated in Figure 7.8, where the crossing of the same train type on two different dates yielded nearly identical measurements.

This data can help identify trains that induce significant changes in signal values. For instance, on one occasion, a passing train caused a noticeable step change in the measured displacement at one of the sensors as shown in Figure 7.9. Further analysis revealed that this train consisted of three locomotives: the first and third had three-axle bogies, while the middle locomotive had two-axle bogies. In contrast, the previous train that had caused the maximum displacement comprised only two locomotives, one with three-axle bogies and the other with two-axle bogies. This comparison suggests that train formation plays a crucial role in influencing bridge stresses.

By comparing the observed maximum values to the standard, it is possible to assess the bridge's margin of safety in operation, as shown in Table 7.1. This table indicates that even under the worst load case, the bridge retains approximately 12% of its capacity.

The results suggested that the dynamic amplification factor for this bridge is 1.27, which is significantly larger than that provided by the standard (1.08). Also,

(a)

(b)

FIGURE 7.6 Installed strain gauges (a) in between blocks in the middle of the span and (b) on the rails to measure the weight of passing trains (WIM).

by updating the numerical model of the bridge, it was possible to have an estimate of the modulus of elasticity of the combined masonry block and mortar. The updated model was then used and it was shown that the bridge could withstand the increased axle load of 22.5 tonnes. This clearly highlights the benefits of using SHM to understand the hidden capacity of the bridges which were constructed based on outdated standards.

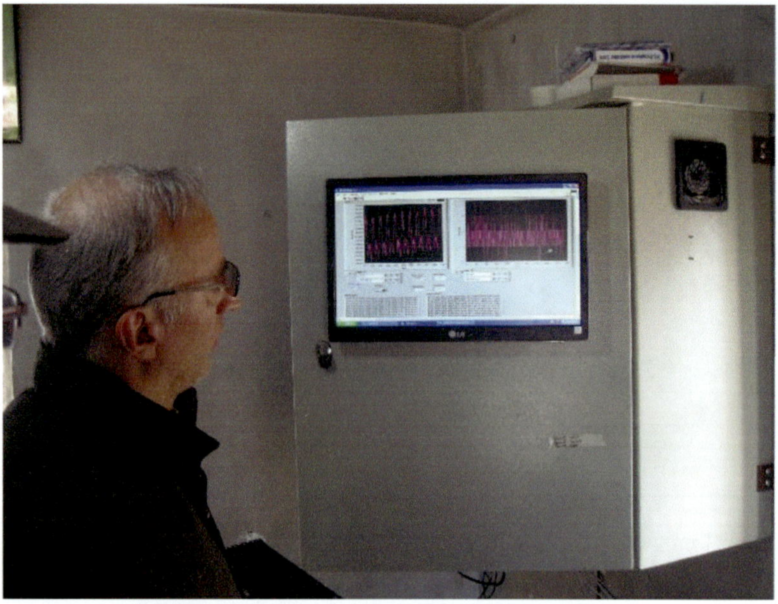

FIGURE 7.7 Veresk Bridge monitoring system under passing traffic.

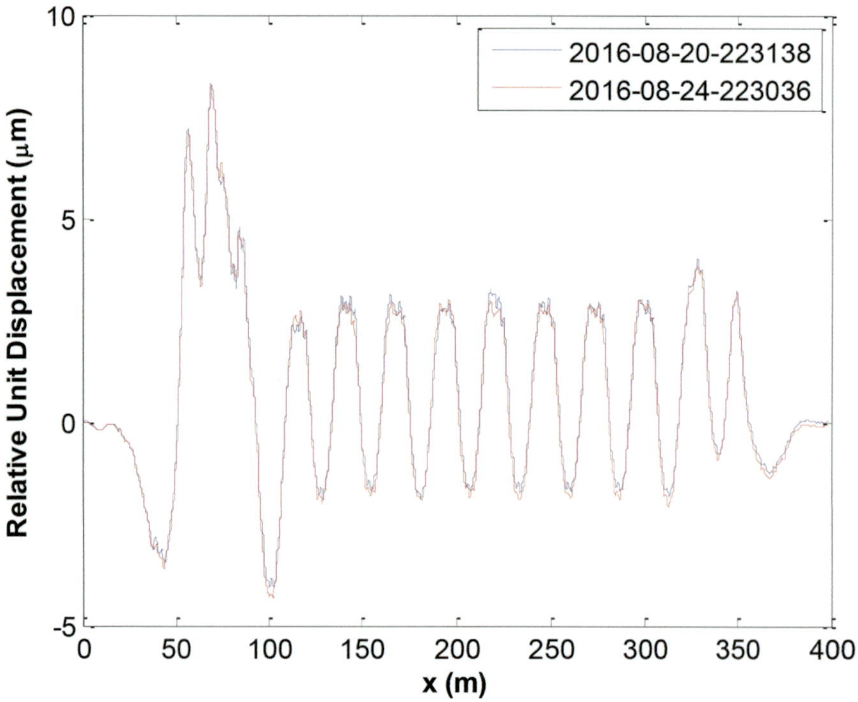

FIGURE 7.8 Relative displacement between two masonry blocks under the same train over two different dates.

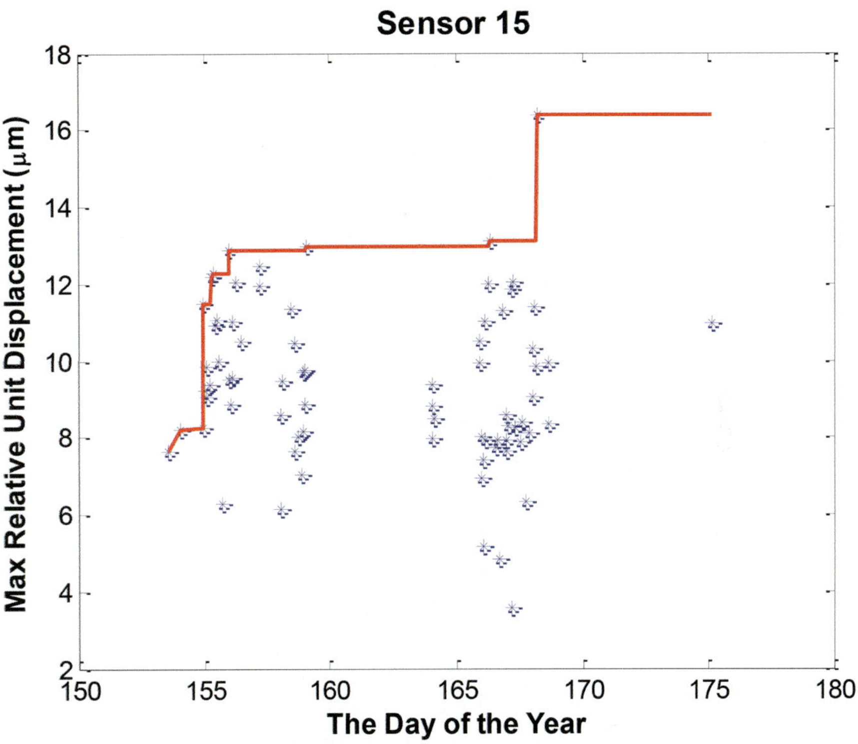

FIGURE 7.9 Significant step change in measured relative displacement values.

TABLE 7.1
Bridge capacity from relative displacement perspective

Sensor	5	3	14	4	6	16	10	15
Allowable displacement	11	30	8	19	15	27	27	27
Measured displacement	7.59	9.80	4.02	16.76	12.39	12.13	10.03	16.37
Measured/allowable	69%	33%	50%	88%	83%	45%	37%	61%

On the demand side, the WIM system was used to get a better understanding of the weight, speed, and number of axles of all the vehicles crossing over the bridge. For example, a year-long monitoring of traffic shows that the 120-ton Hitachi crane, which includes two three-axle bogies with an axle spacing of 155 cm in each bogie and an axle spacing of 146 cm between axles 3 and 4 (the distance between the two bogies), creates the maximum load effects over the bridge, as shown in Figures 7.10 and 7.11. These values can be used to decide on speed restrictions for such loads.

With the installed WIM system on the bridge, it is possible to identify trains that exceed the allowable load limits. For example, Figure 7.12 indicates that 98% of wagons remain within the permitted axle load, with a few exceptions exceeding this

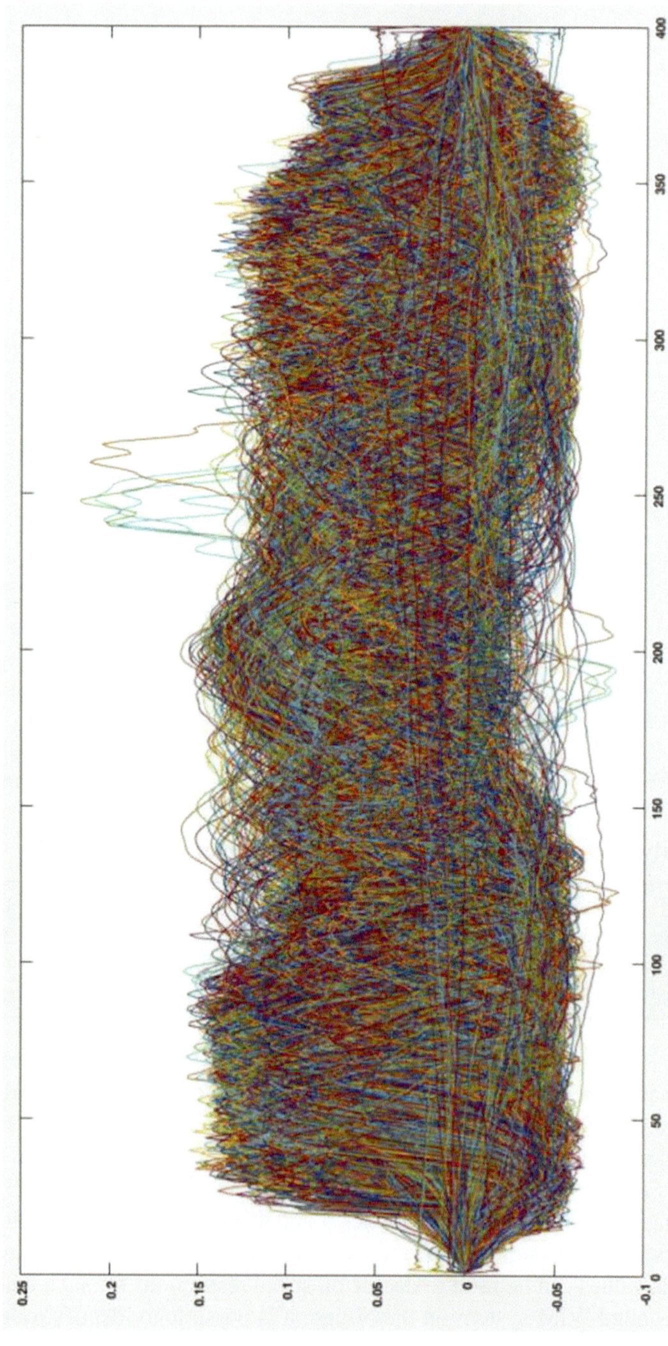

FIGURE 7.10 Time history of the strain at the centre of the Veresk Bridge span in the Veresk Bridge monitoring system under passing traffic over the course of one year.

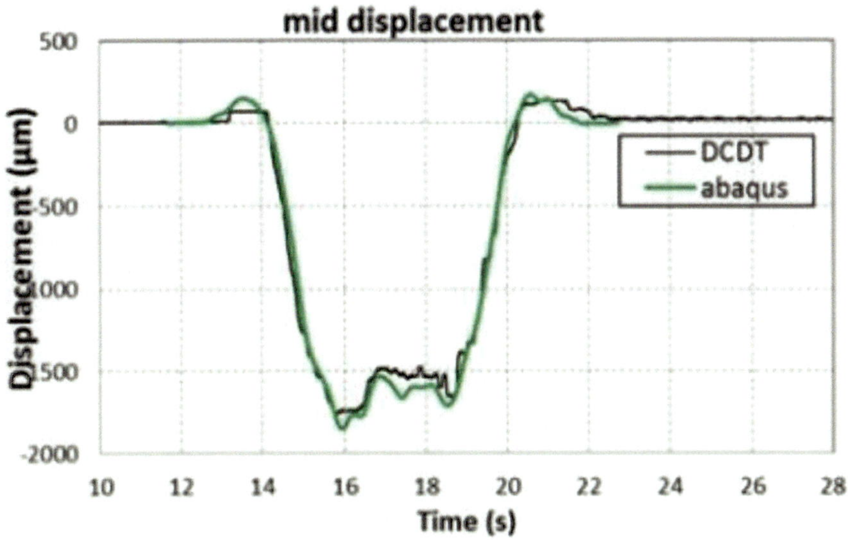

FIGURE 7.11 Vertical displacement at the centre of Veresk Bridge during the passage of a 120-ton double diesel locomotive from north to south.

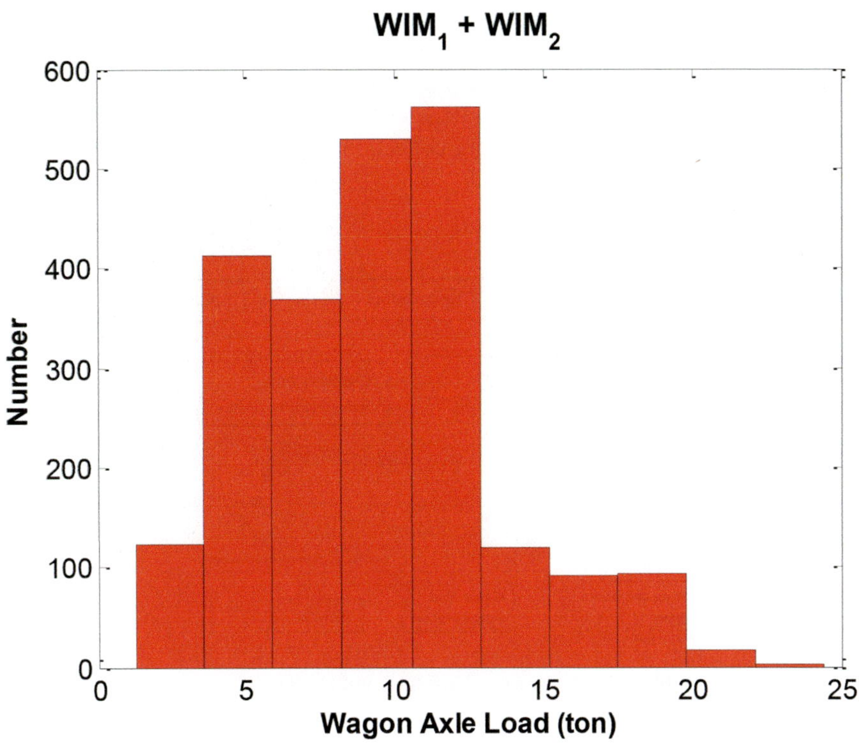

FIGURE 7.12 Axle load of passing wagons on Veresk Bridge.

threshold. This tool provides asset owners with valuable data to support decision-making regarding operators who do not adhere to load or speed restrictions.

7.5.2 Yatri Turnout Monitoring System

Turnouts are vital elements of the rail network because they are responsible for guiding railway traffic in crossings. Turnouts are considered an element that require significant maintenance, given their mechanical design. As such, making decisions that will enhance their behaviour and reduce the required maintenance is of significant importance. Such decisions include reducing operational speeds at through or side crossings, as well as operational axle load on the turnout.

Yatri turnout near Tehran in Iran is located next to one of the major stations and accommodates significant traffic of 40 trains a day with speeds of up to 120 km/h. The turnout is a 1:9 left-hand turnout with a 60E1 full rail section type. Its safe and optimal operation was therefore of significant importance to the asset manager. As such, asset manager decided to better understand the behaviour of this turnout and optimize the speed of the trains that cross over it.

To provide a response to this question, a long-term monitoring system was designed and deployed on this turnout. The SHM system was designed to monitor the displacement of the turnout blade, vibrations in central nose, force on the guiding rail, and strain on the turnout rail. Figure 7.13 shows the instrumentation drawing of the turnout. The system was also capable of measuring the speed and number of axles of passing trains.

Selecting accelerometers with the appropriate measurement capacity is crucial for turnouts, as they experience relatively large vibrations during train crossings. Proper sensor selection ensures accurate data collection and reliable performance under dynamic loading conditions.

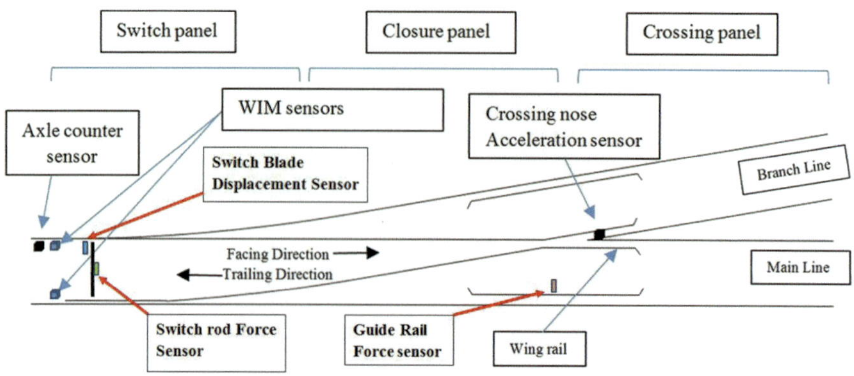

FIGURE 7.13 Installed SHM system on Yatri turnout [4].

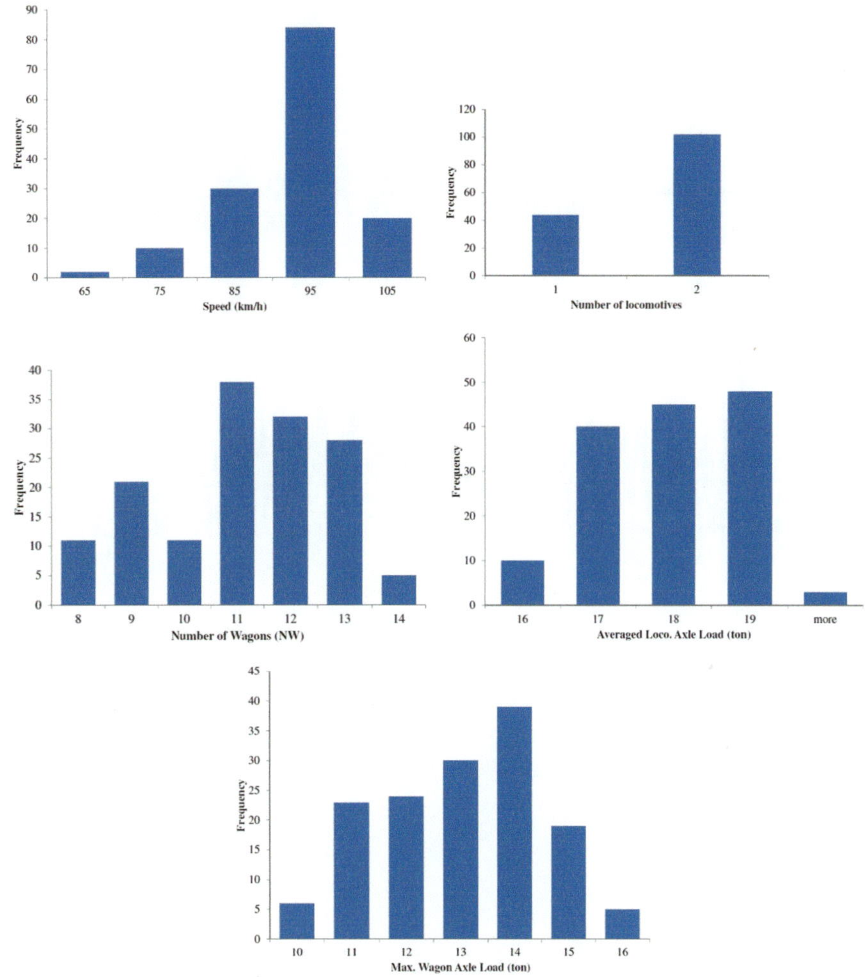

FIGURE 7.14 Train speeds and axle configurations on Yatri turnout over a two-week period [4].

For a two-week period, Figure 7.14 shows the frequency of train formations and speed, along with their axle loads. Also, acceleration signal was processed for all trains and is presented in Figure 7.15.

By training a neural network on the collected data, it was possible to predict the turnout's response under various scenarios. Figure 7.16 presents the results of this analysis. The vertical acceleration RMS limit was set at 95% of the measured accelerations over a two-week period. According to the findings, across all surveyed input parameters, trains reached this limit at speeds of approximately 100–105 km/h in the most severe cases. This information enables asset managers to make informed decisions about balancing increased operational speeds through the turnout against the associated maintenance costs.

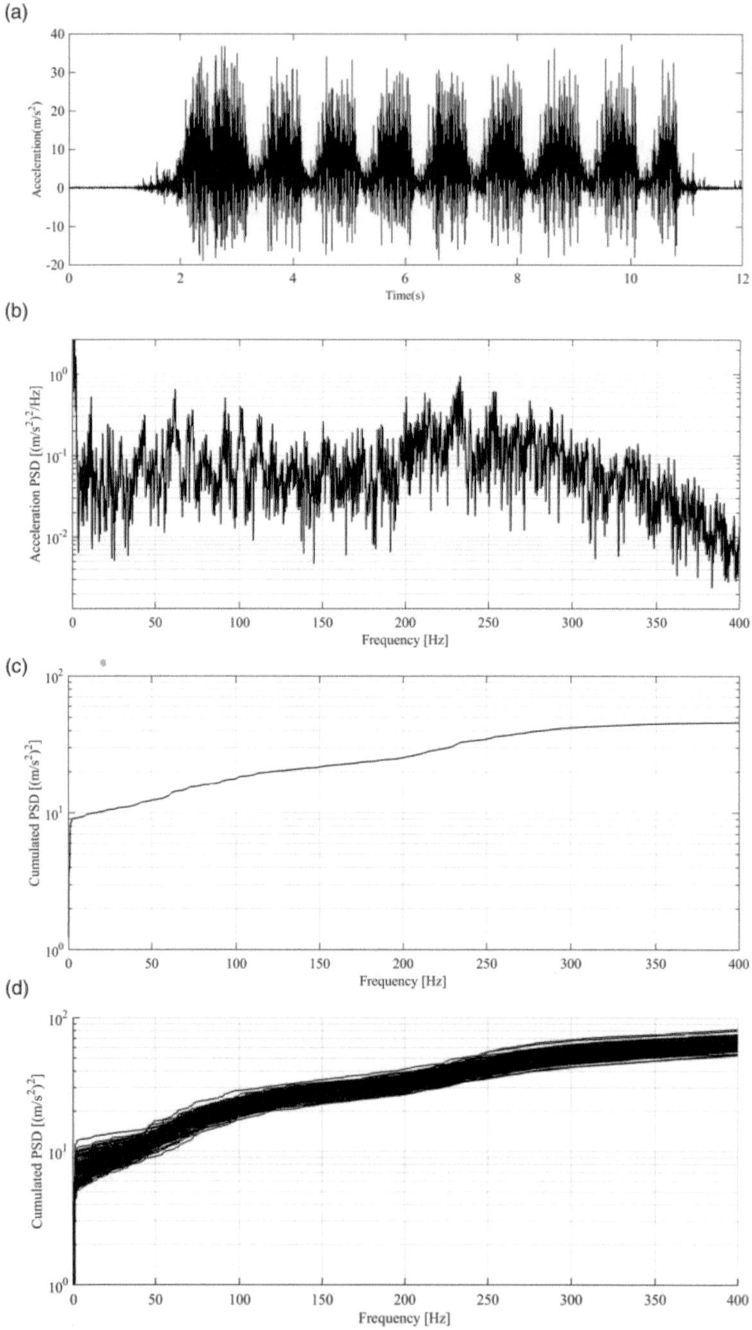

FIGURE 7.15 Observed accelerations on Yatri turnout over a two-week period [4].

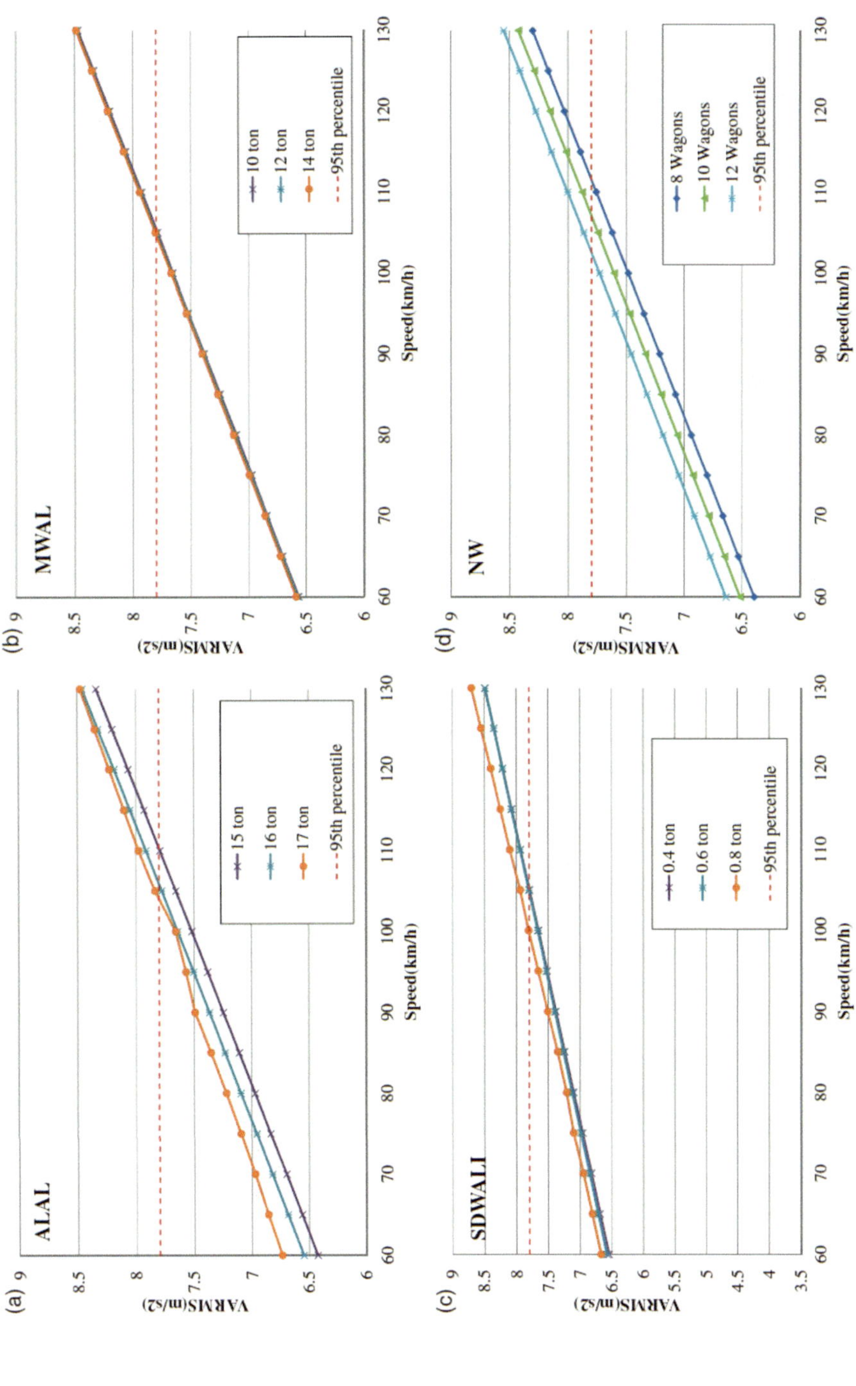

FIGURE 7.16 Sensitivity analysis for (a) average locomotive axle load, (b) maximum wagon axle load, (c) standard deviation of wagon axle load, and (d) number of wagons [4].

7.6 IDENTIFICATION OF OPERATING LOADS

Train weighing systems are critical components in the railway infrastructure. By measuring the load of trains as they pass over tracks, these systems ensure compliance with regulatory weight limits, thereby mitigating risks associated with overloading. Overloaded trains can lead to significant damage to railway infrastructure, including track deformation, increased wear on rails, and even catastrophic failures such as derailments. Regular monitoring through weigh-in-motion (WIM) systems helps to maintain the structural integrity of both tracks and rolling stock, extending the lifespan of railway assets and optimizing maintenance scheduling.

A WIM system is a widely adopted technology that weighs trains at regular operating speeds, enabling dynamic monitoring that does not interrupt traffic flow. These systems are equipped with sensors installed either on or under the rails. As the train moves over the sensors, they measure the vertical forces exerted by each wheel and axle, calculating the weight distribution of the train. This data is transmitted to a central processing unit that analyses the load against predefined safety thresholds. In addition to weight monitoring, other parameters like speed, axle spacing, and vibration data can be collected and analysed to gain a holistic view of a train's interaction with the track infrastructure.

The resulting data can provide a first-hand view of the types of loads that cross over the network. Using this data, it is possible to assess the condition of any structure not necessarily based on standards loads, but actual operational loads. Naturally, this provides a more accurate estimate of the condition of the bridge. It should be noted that standards have many assumptions, which result in conservative loads that are not always reflective of the actual operational loads of the bridge, and hence a structure may not be compliant with such loads but have been in service for many years without any issues. This is particularly the case for older structures that were constructed based on older standards of the time.

7.6.1 Iranian Railways WIM System

A similar system has been developed in Iran, details of which are shown in Figures 7.17–7.20. This system consists of four layers: design, measurement and collection, transmission, and data analysis and management.

- The design layer is created with the goal of weighing the axle load with 1% accuracy for speeds over 15 km/h.
- In the measurement and collection layer, based on the objective set in the design layer, strain gauges are used to measure the shear changes between two sleepers, and 32 sensors are fabricated, installed, calibrated, and operated.
- In the transmission layer, all the collected data from each train crossing is transmitted to the control panel.
- In the data analysis and management layer, after preprocessing and analysis, and separating environmental factors (temperature) and operational factors (speed), the wheel weight is determined and reported for management and monitoring purposes.

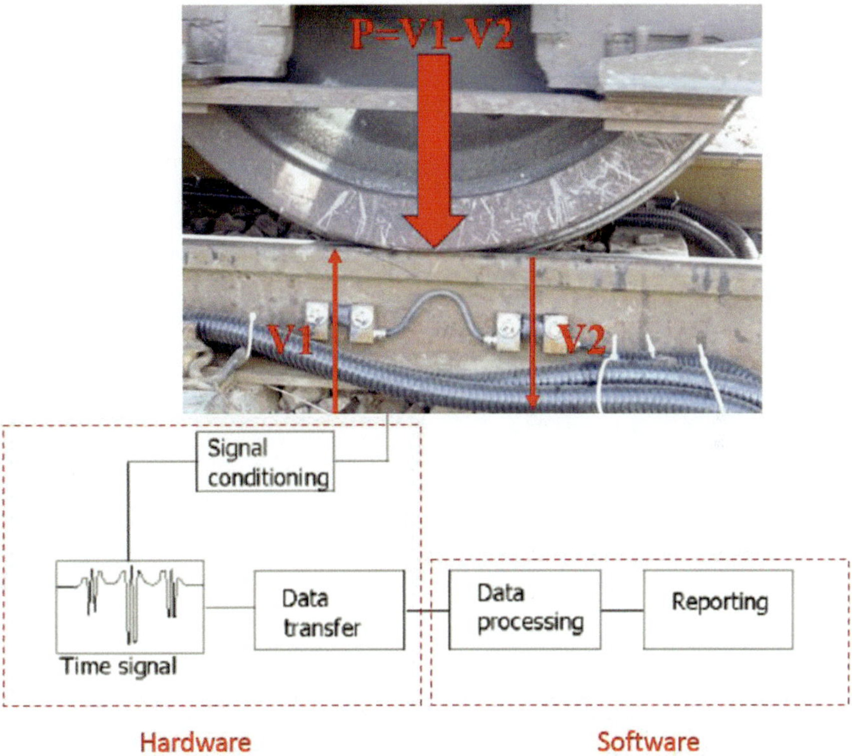

FIGURE 7.17 Dynamic axle load weighing system of the train.

FIGURE 7.18 Dynamic axle load measurement of the train based on the AMTAB system in the Azerbaijan region of the Iranian railway.

FIGURE 7.19 Measurement of the axle load of a moving train wheel based on shear variation measurements.

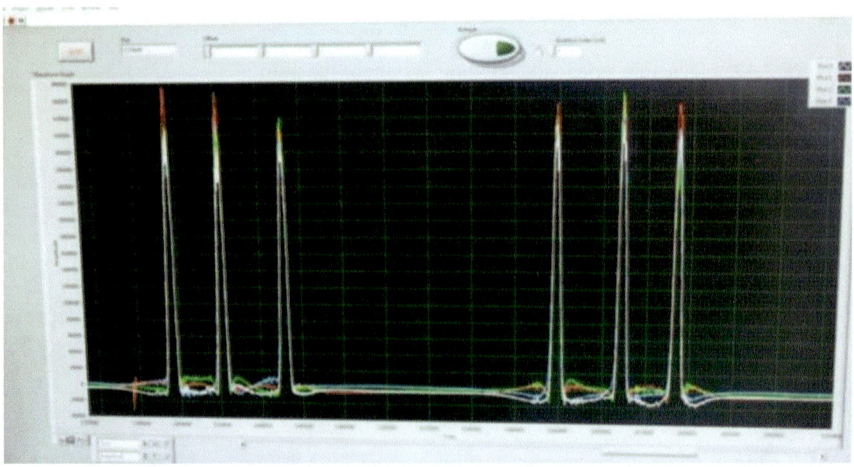

FIGURE 7.20 Measurement of the moving six-axle diesel locomotive axle load.

A unique feature of this system is that it does not require concrete substructure, rail welding, or cutting, is fully modular with no moving parts, can be installed without interfering with passing traffic, weighs without affecting operational speed, does not require specialized personnel for maintenance, and can send data to the railway network and create alerts to prevent unauthorized axle loading.

The software for the moving train axle load measurement system calculates and reports the following decision-making indicators:

- Calculation of axle load
- Simultaneous display of the passing load graph
- Calculation of the load of each wagon separately and the total load of the train
- Calculation of train speed, the number of wagons, and axles
- Display of the left and right wheel weight of each axle and calculation of the asymmetric load distribution on each axle.

7.7 SUMMARY

In this chapter, service life planning of a structure was discussed. It was explained how Maintenance Management Excellence Model, which applies ISO 55000 principles to maintenance by implementing structured, data-driven approaches can be utilized to improve the service life planning of assets. The final goal is to adopt maintenance measures that can extend the life of the asset.

The chapter explained the role of SHM in achieving maintenance excellence model. In general, structural capacity must exceed the demand induced by code-specified loads. However, in real-world conditions, this relationship may not always hold for constructed structures. Various factors, including material variability, construction tolerances, ageing, and unforeseen environmental influences, can lead to deviations from design expectations.

Despite these uncertainties, structures often exhibit hidden capacity, meaning their actual strength and resistance exceed the theoretical values determined by conventional design methods. This hidden capacity is due to factors such as load redistribution, reserve strength in materials, and unaccounted structural redundancies. Similarly, the loads acting on a structure are sometimes lower than the conservative estimates provided by design standards. Traditional codes accounted for uncertainties by applying safety factors, which, while necessary for general reliability, occasionally result in an overestimation of required capacity.

The role of SHM in understanding the extent of hidden capacity and actual load conditions were discussed through practical examples. On one side, by continuously monitoring the assets, it is possible to understand the overall response of the structure to imposed loads. On the other hand, measuring the actual loads will assist in reducing the uncertainties considered in loadings in standards, therefore allowing for the structure to have a higher capacity than that determined through a standard.

By integrating SHM data into structural assessments, engineers can move towards performance-based maintenance, ensuring that structures are not prematurely decommissioned due to excessive conservatism in standards. Also, SHM data can be used to adjust maintenance scheduling in case the operational loadings are higher than those assumed during design and operation assessment. This approach supports sustainable infrastructure management by extending service life and optimizing resource allocation.

REFERENCES

[1] J. Moubray, *Reliability-Centered Maintenance*. Industrial Press Inc., 1997.

[2] D. M. Frangopol and M. Liu, "Maintenance and Management of Civil Infrastructure Based on Condition, Safety, Optimization, and Life-cycle Cost," *Structure and Infrastructure Engineering*, vol. 3, no. 1, pp. 96–106, 2007.

[3] "ISO 15686-1, Buildings and Constructed Assets, Service Life Planning," 2011.

[4] K. Mehrzad and S. Ataei, "Railway Crossing Vertical Vibration Response Prediction Using a Data-driven Neuro-fuzzy Model—Influence of Train Factors", *Journal of Rail and Rapid Transit*, vol. 235, pp. 1086–1098, 2021.

[5] S. Ataei, M. Nouri and V. Kazemiashtiani, "Long-term monitoring of relative displacements at the keystone of a masonry arch bridge", Structural Control and Health Monitoring, first published 19 January 2018, doi:10.1002/stc.2144

Index

For Product Safety Concerns and Information please contact our
EU representative GPSR@taylorandfrancis.com Taylor & Francis
Verlag GmbH, Kaufingerstraße 24, 80331 München, Germany